Patricia Gosling is a retired psychoanalytical psychotherapist in the British Independent tradition. She has a background in Medicine, Marriage Guidance, Clinical Theology and the hurly-burly of family life. Patricia has been a member of the Religious Society of Friends (Quakers) for nearly fifty years.

Other books by Patricia Gosling:
A Curious Eye (2006)
Loving and Loss (2009)
A Time of Transition (2011)

The front cover shows a new-born infant, the back a scene of devastation from the Spanish Civil War (1936-39). Both are superimposed on the Milky Way, our galactic home.

Fatal Flaws
Notes from a perverse society

Patricia M Gosling

WHITE HART BOOKS

© Patricia Gosling 2012

The right of Patricia Gosling to be identified as the author of this work has been asserted by her in accordance with the Copyright, Designs and Patents Act 1988.

All rights reserved

ISBN 978-1-4710-7820-0

No part of this work may be reproduced, stored in a retrieval system, or transmitted in any form or by any means, electronic, mechanical, photocopying or otherwise without the prior permission of the copyright owner.

Published by
White Hart Books
Rode, England
hart@gosmob.eu

To:
those many angels unaware
who have sustained me
on the journey

Contents

Introduction 5

Acknowledgement 8

1 Fatal flaws 9

The illusion of perfectibility 11
Me / Not me 15
The lust for power 19
Slavery 19

2 The myth of inexhaustible supplies 23

The roots of financial incompetence 25
Can we learn? 27

3 The elephant in the room 32

The population problem 33
The crisis for men 35

4 Depression in men 38

Where have all the fathers gone? 40

5 The politics of gender 42

The courage of women 43
Kept women? 45

6 The crisis of authority 47

The origins of violence 48
Containing aggression 49
Practical politics 52

7 Post-imperial blues 54

8 Socialism 58
>*Bureaucratic centralism 60*

9 Democracy 64
>*The need for autonomy 65*
>*Open loop, closed loop 66*

10 Social identity 68

11 What is out of control? 72

12 Twentieth century blues 75

13 The perverse society 83
>*Perverse organisations 86*

14 The perverse society in terminal throes 90

15 Where next? 96

16 New world, new language 105
>*The limitations of words 105*

17 Contemporary religion 109
>*Sure and certain hope? 112*

18 Psychotherapy and spirituality 115

19 In the manner of Friends 124
>*The Peace witness 125*
>*A military culture 127*
>*The rise of feminism 129*
>*Quaker pacifism 130*
>*Aggression and the religious life 131*
>*Barriers: to break down or respect 133*

20 Quaker mysticism 137
>*Experiences of the numinous 138*
>*The Inner Light 141*

21 Life has a purpose 143
Psychoanalysis and the Bolsheviks 144
Beyond the pleasure principle 148

22 Religion in a post-modern world 152
What are the challenges of our time? 155

23 Credo 159

24 The journey toward P6 163

25 Riding the waves 167

Introduction

As I have gradually moved into retirement, and my preoccupation with the internal world has become less compulsive, I have become more aware of — and more interested in — what is going on in the world around me. This book is the result.

I have come to realise that I am someone who knows what she thinks only when she finds herself writing it. The result has often surprised me. Now that my professional persona no longer demands a non-judgmental stance, and a willingness to accept whatever comes my way, I find I have a lot of powerful feelings about many issues. Expressing those feelings in writing has been distinctly therapeutic, and protects those around me from the boredom of listening to my repeated tirades.

I have become more aware that — inevitably — the world has changed a great deal since I was young, and I now have the thinking space to put those changes into perspective. It seems important — not least so that one does not burden the young with advice which is no longer relevant.

Having lived a long life enables one to have some sense of those barely perceptible but hugely significant changes that occur in any society over time. In their different ways the Pill and the Inter-net have had profound influence. Longevity also helps gain perspective on the changing fashions of thinking which too readily trap us all into false certainties.

The early chapters are essentially descriptive of our contemporary society and culture as I have experienced it. In chapters 7-9 I attempt a theoretical framework. It is one which is formed out of my psychoanalytical perspective, and which has enabled me to give some coherent meaning to events.

The underlying theme of this book — unspoken but omnipresent — is existential anxiety. The species of Homo sapiens sapiens has great intelligence, and with that a high degree of self-consciousness. The downside of this is an uncomfortable awareness of our vulnerability and dependency.

I believe that the devastating events of the twentieth century, and the launching of the two atomic bombs in 1945, have made us acutely aware of our powers of destruction, and have frightened us. I believe that what I have described as the perverse society was an attempt to deal with this fear by denial, and hypomanic activity. That has now broken down.

An important function — perhaps *the* most important function of religion — is to provide a framework of thinking, of living, which helps us to contain and manage that existential anxiety. Otherwise it threatens to paralyse and overwhelm us.

I have come to realise that my religious faith, such as it is, is central to who and what I am.

I have spent many years as a practising Quaker, and I much value that tradition. Subtly, and over a long period of time, it has formed me, and helped to shape my perspective on the world, my Weltanshauung. It has provided me with a community of sufficiently like-minded people to offset the loneliness of being a vulnerable adult in a difficult world; and particular individuals have been an enormous support at times of stress. This experience is relevant to the later chapters which touch on spiritual matters.

I suspect one needs a certain kind of temperament to be a member of the Society of Friends, and I accept that others may need something quite different in the way of religious practice. Whatever — I believe our world has a great need for what religion can offer. Finding a formulation which suits us temperamentally, and which is consonant with our complex and rapidly changing world is no easy matter. It is a task we each need to address. It is a task for the years ahead.

Since completing this manuscript, I have come across my old battered copy of anthropologist Robert Ardrey's book *The Social Contract*, first read in 1972. I realise that in it he pre-figured some of my own ideas and perspectives, notably his insistence on biological reality and the folly of ignoring it.

I would like to think that such ideas do go on quietly fermenting in the recesses of readers' minds, since that is the only possible excuse for the activity and effort of the writer.

Acknowledgement
The ideas in this book were originally written in the form of observations, spontaneous outbursts, and occasional essays, usually stimulated by happenings in the outside world. I am grateful to William Gosling for his efforts in melding them into something approaching a coherent narrative - and for his evocative cover design.

1 Fatal flaws

The history of the human race is an incredible evolutionary success story. From small beginnings this naked ape has populated the surface of the earth, acquiring on his journey the skills of living in a wide range of environments. Earlier varieties of humanoid have faded into oblivion. Many other species of life have also disappeared, or are currently in process of vanishing under the competitive pressure on their habitats.

Homo sapiens sapiens has learned to fly in the air and to travel under the sea. He has used his intelligence and resourcefulness to make permanent shelters, to grow and farm his food, to extract ores from the earth, to forge tools and weapons, to build cities and cultures, to create civilisations and empires. His ingenuity and inventiveness seem inexhaustible.

And yet....and yet.... We do not have paradise on earth. Warfare is endemic; intermittently it flares acutely, and we talk of 'war to end wars', or 'ousting an evil regime'. In reality, after the acute phase is over, warfare again becomes sporadic and chronic. Ethnic cleansing too is endemic. The name may be new but the process is not, and once begun readily escalates into attempted genocide.

Throughout known history, the political process has allowed power into the hands of people who were deeply disturbed, pathological and immensely destructive to those in their ambit. One thinks of Tamburlaine, of Ivan Groszny (the Terrible), of Henry the Eighth of England in his latter days, of Napoleon Bonaparte. The 20th century spewed forth a clutch of such monsters — Hitler, Stalin, Pol Pot, Mao Tse-tung

Even for those of our young not currently engaged in warfare, all is far from well. In the more affluent parts of the globe, drug abuse is widespread, made more dangerous, in the attempt to contain it, by criminalisation. Suicide is disturbingly high amongst young men.

Many of those who come through to stable adulthood decide not to procreate — arguably no bad thing. More worryingly, many of those who do have children cannot or will not devote the time and energy necessary to adequately care for their young, nor are they willing to stay with the extended process demanded by the protracted length of human maturation. The skills of parenting are, to a considerable degree. learned unconsciously from one's own experience of being parented. In dysfunctional families, these skills can all too quickly be lost.

So much in our contemporary society is currently weighted against good parenting, from the cost of housing to fashionable ideas which bear little relation to the reality of family life. The wonder is that so many do it as well as they do. Winnicott suggested that a democracy needs a minimum of 10% of mature, stable people to function properly. Are we producing that minimum? If not, we are laying up long-term problems for ourselves.

I want to suggest that human beings have certain inbuilt tendencies which are currently fouling up our social order. If we cannot gain some insight into them, and with the insight some capacity to moderate them, then all our attempts at 'solutions' are doomed to failure. In the (very) long term what is at stake is not just the future of our world and our children's and grand-children's world, but the future of our species.

Earlier versions of mankind have become extinct. There is no law of nature that says it cannot happen to us.

The illusion of perfectibility

It was Jung who coined the word archetype to describe some innate constellations in the human psyche. Although popular amongst his followers, his ideas were somewhat marginalised by the more 'scientific' formulations of Freud and his school of psychodynamic theorists. It is only recently, some hundred years later, that the concept has been given new life from a very different source.

The research conclusions emanating from neuroscience suggests that the human brain at the time of birth has within it patterns of organisation which give us certain propensities of behaviour. These patterns are what is known as *hard-wired*, that is they are a function of the physical state of the brain, only to be lost by physical damage — as opposed to *soft-wired*, those patterns acquired by subsequent experience which theoretically can be modified by learning techniques.

This hard-wiring is inherited by the same genetic and other mechanisms which tell our developing foetal bodies which organs and limbs to grow at which stage of intra-uterine life. At a later stage of life, it tells our bodies when to undergo the hormonal changes leading to puberty, then to menopause in women, and subsequently to death. It monitors our incarnate existence.

It is this innate patterning which allows us to learn language so young and so quickly. It is also this kind of patterning which drives our appetites, our instinctual life. It is as if at some level we know what we need and are driven to try to find it.

I have come to believe that we all have a sense of our own innate potential, and are all the time attempting to actualise it — trying to find those people, objects and environments in the outside world which will enable us to develop to our full capacity. It is perhaps a sign of maturity when we come to accept that we do not live in a perfect world, and that some of our potential never will be realised, indeed cannot be, since whenever we take one decision, we foreclose on other options

The Christian might talk of 'a fallen world' or 'original sin'. The psychotherapist might talk of proper disillusionment (*pace* Winnicott) and the achievement of the depressive position. It is something about coming to terms with the reality of one's life situation, while continuing to hope and to strive for the best one is, and the best one can do. It is a tightrope which enables us to be both realistic and creative.

It is a help if one has investment in the younger generation, whether of one's own family or others. By facilitating their development, one can alleviate the disappointments of one's own life. Erikson has something to say about this in his concept of 'generativity'.

This innate tension between what at a deep level we know we have it in us to be, and our failure to potentiate it, stays with us. It spurs us to achieve. It is a source of that 'divine discontent' which will not allow us to rest on our laurels for long. It can be both a force for good or a force for ill. It can evoke creativity of all kinds: it can turn us into egocentric monsters. It can drive us to live life to the full: it can leave us bitter at our failure. So much depends upon how this innate drive is handled in our formative years. If it is seen and met and responded to positively, we learn from those significant others who we are and what we might be. Sadly, much failure and subsequent regret is initially a failure of the facilitating environment to recognise and support the child. Sometimes, happily, the individual finds mentors later in life who give him a sense of who he is, and of his own worth as a person.

Living with tension is never comfortable. One way to relieve this tension is to hang on to fantasy, to refuse to accept the inevitable compromises and letting-go, the minor bereavements, the 'proper disillusion' that is part of the depressive position of healthy growing-up. [1]

The notion of perfection and perfectibility is immensely seductive. It is a clinging-on to that infantile sense of vast potential, of omnipotence, of worlds waiting to be conquered, of heroes and gods and fabulous wealth and impossibly beautiful beings. It is a clinging-on to the internal image of the ideal parent who can give one everything one asks for, and never lets one down. It is a refusal to mourn.

One sees it in full force in the manic phase of bi-polar disorder. One sees it, no less, in the current obsession with the perfect body, the perfect home, the perfect life-style. One sees it getting out of control in mounting personal debt, in anorexia nervosa, in that slogan of Gestalt Theory that 'we can become anything we want to be'.

Not true! - and in that denial of reality lies a denial of our biological roots. Yes, we all have undeveloped potential.

---

[1] The concept of the depressive position was first propounded by Melanie Klein (1882-1960) to describe that phase of an infant's life when it begins to develop an awareness of the real world outside itself.. It comes to realise that the mother it loves and the mother it hates are one and the same. It is the beginning of the capacity to mourn, to feel concern, and to wish to make reparation.

However, it is also true that we each have our limitations, both innate and circumstantial. The skill of living a satisfying life lies in using the transitional space between the two.

Both the individual urge to become perfect, and its intellectual expression in the belief of man's perfectibility, has a long history. Plato believed that behind this world lay an Ideal world. Many religions have a concept of Heaven — the perfect place where man can achieve a perfect state of being. The Enlightenment was based on a belief that we could make 'Heaven on Earth', as were the more recent doctrines of Socialism and Marxism. Utopias have abounded from Sir Thomas More's version to Samuel Butler's *Erewhon* to William Morris's brotherhood.

Christianity has been less than helpful at times, feeding into our illusions of perfectibility. The mistranslation of Jesus' words into "Be ye perfect" is now understood as better rendered "Be ye whole" — which has a different flavour. The Church's attempts to coerce its adherents into proper behaviour by the dubious and improbable examples of sainthood has done a lot of damage. (Even where the tales are substantially true, the historical saints were often anything but 'saintly' in their behaviour, and frequently come over as distinctly awkward people.)

To quote Rowan Williams[1] 'The Christian Gospel tells us not simply that we are saved from our sin or that our guilt is taken away — it insists that we find out who we are and what we may be in an encounter, a relationship...' (Williams 2000) This is no easy phrase, pandering to our fantasies of perfectibility, but the words of a man who understands the theological (and costly) doctrine of Incarnation, of being a human being in this world.

---

1 Appointed Archbishop of Canterbury in 2003

The only perfect world we humans have ever known has been the womb. It is perhaps a dim memory of our intra-uterine life which fuels our concepts of Heaven and the blissful life to be found there. Similarly, our sense of ourselves as perfect beings perhaps also has its roots in that time when we were, just by being ourselves, without effort or struggle, just perfect. The Genesis story said it all. With self-consciousness came expulsion from the Garden of Eden, and the command to work out our salvation.

Me / Not me

We've all met them — those people for whom it is always someone else's fault. A student about to be expelled from college says, 'You're getting rid of me because I'm black and you're prejudiced.' The response was, 'No, it's because you are not coping with the work.'[1]

A woman with several children ran off with her lodger. 'It wasn't my fault,' she said. 'He persuaded me.' At one level it was not her fault. Allowing herself to be seduced was a pattern in her life, and the only way she knew to obtain the love she craved. (She had been sexually abused by her adoptive father.) However, since she was a decade older than the young man involved, she did not receive much sympathy.

Other cases are not so clear. A couple begin to have sexual intercourse, then at some point the girl has misgivings and cries rape. Our legal system has difficulties with this one, as it does in cases where an employee from a racial minority gets the sack. The law feels to be too blunt an instrument in such instances..

When a toddler still barely in control of his limbs falls over a chair, he says 'naughty chair' and we laugh and soothe him. We know it is his clumsiness, but we allow him the small distortion of truth in order to preserve his courage

---

[1] The student was allowed to repeat the year, and failed again.

and self-esteem. That is fine for toddlers, but when we never progress beyond that stage of development, we are in trouble. Marriage, intimate relationships of all kinds, tend to evoke the primitive aspects of our personalities. It is because such relationships provide a safe place for those bits of us that we crave them. However, when they begin to break down, the quarrels can be horrendous as each party accuses the other of being in the wrong. 'It's all your fault!' When whole communities, whole countries fall into this mode, then we are in serious trouble.

Narcissism is normal for the human infant. So is disillusion. None of us like our bubbles to be burst — it hurts. Psychotherapists talk of a baby's eighth month depression. It may sound a ludicrous concept, but it is real. At about this time the child begins to realise that its mother is a separate person and is not under his/her total control, and he or she doesn't like it one little bit. It grizzles and complains and makes its unhappiness felt. If it is fortunate in its care-giver, it will be nursed and spoilt a bit and encouraged and not left to flounder too deeply in its (very real) misery. Before long life will seem worthwhile again, if never quite what it was. An important hurdle will have been surmounted.

However, the child will have learned a difficult lesson — that what is *not me* can hurt. That kind of lesson takes some assimilating. It is a lesson that will be repeated over and over again in a myriad of small ways, and against resistance. One form of resistance is to push the pain away: 'It doesn't belong to me. It is someone else's fault. It is someone else doing it to me', (which may or may not be true). It takes a lot of maturity before a person can say: 'This hurts, but it is not anyone's fault. It is just the way things are.' Much easier to declare: 'Nasty stone for hurting my toe; hateful Mummy who won't let me do what I want.' It is all too easy to go from there to: 'Nobody loves me, they are all against me, its all their fault.'

We are talking here, of course, of the mechanism of projection by which the child (and the child within us) protects itself against what is too painful to cope with — the loss of self-esteem and the loss of a sense of well-being. One hopes that the adult has sufficient resilience and capacity for thought to recover emotional balance before too long. The child has not yet acquired those skills.

Some of us never do. We continue to persecute those who pain us by their actions, by their difference, by their very existence. It makes us punitive to those who offend us, even when rationally we are told that such punishment exacerbates the situation. We treat our social criminals in a way which confirms their criminality. We try to distance ourselves from those whose difference we cannot cope with. We try to expel from our territory those whose customs feel strange to us. Before we know where we are we are into anti-Semitism, Islamophobia, ethnic cleansing, the Holocaust and such-like atrocious scenarios.

We are addressing a number of highly important issues here. An immutable fact is our extreme vulnerability at birth compared with other mammals, We need a great deal of care, both physically and psychologically. We are, to begin with, totally helpless and dependent on those around us. If we are fortunate in our care-givers, this reality does not hit home except occasionally, which is as well because that reality is terrifying. The mechanism of projection is one of the ways we protect ourselves from what could otherwise be overwhelming and disintegrating fear.

Our society has become very good at understanding and providing the necessary physical child-care. Most of our infants survive. Peri-natal and post-natal mortality figures have dropped steadily and the outlook is good. We are not so good in providing the requisite emotional care; and a succession of unfortunate social fashions have tended to undermine women's instinctive understanding of their infants' needs.

In the recent past we have had the Truby King era of rigid 4-hourly feeding and minimal physical contact. We have had the separation in hospital of mother and newly-born infant at a time when the bonding process is paramount. We have had a feminist perspective which expects women to work throughout pregnancy and to return to work as soon afterwards as can be managed, leaving child-care inadequately addressed. We have all those social pressures which have led to an increase in the number of single mothers; and recently we have the necessity to work to meet the scandalous cost of housing. We have addressed the issues of physical care well. We all too often do not want to know about the concerns around emotional care.

The price is high either way.. Children who do not get the care they need at the time they need it, tend to demand attention later in all manner of time-consuming and tiresome ways. We have all met the child who always wants, wants, wants — treats, toys, outings, soft drinks, sweets — and is never satisfied. We are rearing children so unsocialised that their schools cannot cope with them, and with a higher level of physical aggression than their elders can recall. We are rearing too many children who go on to become addicted to drugs.

It is actually easier for all concerned if the infant is given the emotional care it needs at the time it needs it. Yes, it is inconvenient, disruptive, demanding, exhausting, It can also be deeply satisfying. It should be seen as an investment. Only if a child feels loved can it love itself, and learn to love others. Material things are no substitute

The lust for power

The underlying psychodynamics of the common human urge to exercise power are not difficult to understand. As infants we are literally powerless. During the long years of childhood, we are under the control of the adults around us, and the only power available to us is to make ourselves a nuisance to those adults.

If we are fortunate, our powerlessness is tempered by the love and concern of the powerful adults. If not, we are left with an agenda which can range from a dislike and distrust of authority, to a determination never to let ourselves be at the mercy of others ever again. If there has been real neglect or abuse, we may identify with the aggressors, and be left with an urge to visit on others what was done to us.

There are those who retreat from relating to people into a world of things — machines — over which one can have a large measure of control. Others may become addicted to the virtual worlds of computers and the inter-net. Yet others retreat even further into a world of ideas. If they are highly intelligent, they may become mathematicians or philosophers.

A threatening solution is that which develops into a need to control other people. On a personal scale, it can wreck relationships; in the social sphere, it is potentially dangerous. A democracy has inbuilt safeguards against those politicians who aim to control everything and everybody, but in a less developed society the threat of dictatorship is all too real.

Slavery

The institution of slavery is another social expression of a need to have total control over others. Slavery is heinous — immoral and degrading. We deplore our historic involvement in the trade, and revere those, such as William Wilberforce, who worked to abolish it.

But slavery was not just a phenomenon of a limited historical era, roughly 16th century to 19th century. The Hebrews of the Old Testament had slavery. Virtually all the ancient civilisations had it. The Greek civilisation was built on slavery, as was the Roman. We admire the Hellenic achievement of democracy, but in classical times it applied only to free men. In Athens at its peak, it is estimated that each household had an average of four slaves, while 25% of the inhabitants of Republican Rome were slaves.

The discovery of a slave chain in Llyn Cerrig Bach in Wales suggests that the Celts were not above the practice. St. Patrick of Ireland himself was a Welsh man taken slave as a boy by Irish raiders on the north-west British coast.

Slavery was endemic in Africa long before European ships beached on African shores. The vendors were Africans from other tribes, while Arabs from the north were their export market. It was a lucrative trade, and as these things go, the slaves of Arab masters were not treated too badly.

Europeans along the Mediterranean coasts told of Arab pirates kidnapping their women for the harem[1]. Arabs will tell you that the women were bought fair and square from their families with good money. Who knows the truth in reality? The girls may have had less impoverished lives than if they had not been traded. A few rose to positions of great influence and power, and were celebrated. Today we are also seeing trade in woman-flesh, some for domestic work, some as sexual objects, from eastern Europe, from

---

1 Islam sanctioned limited polygamy at a time when male life expectation was very low, due mainly to warfare. However the new religion also enjoined a more peaceable life-style. The two factors together created a chronic shortage of women.

Asia, from Africa. Numbers are uncertain. It is a repellent trade.

Good liberal opinion and effort notwithstanding, it looks as if slavery is an institution that is always with us. Why? What is the essence of slavery? The answer seems to be total control over another human being. Someone who will meet one's needs, submit to one's desires, do the dirty work of life without complaint, without reward, without demanding the mutuality of a truly human relationship. Slaves are depersonalised, functional and disposable.

We all have slaves at a certain stage of our lives. They are called mothers, (or nannies, ayahs, grannies or some such.) Sometimes we have several at our beck and call. Without them we should never survive the first few months of life. However, the prolonged persistence of that mode of relating produces selfish, narcissistic monsters. Who was it who said 'absolute power corrupts absolutely'?[1]

Frustration is a necessary learning experience. We need support in order to learn to tolerate it. Without that help we do not develop the mechanisms of containing our unease and distress, and are all too willing to project it into others. Slaves are the ideal recipients of such projections. They cannot complain since they are not fully human. There is no real recognition that they too hurt.

Slavery, at best, is the institutionalisation of immaturity; at worst, it is sadism. It is the substitution of power for mutuality and, as with all perversions, with acting-out it tends to escalate.

We are fortunate to live in a society which is wealthy enough and ingenious enough to have developed machines which relieve much of the drudgery of living. If they infuriate us by their complexity or unreliability, we

---

[1] Lord Acton (1834-1902). 'Power tends to corrupt, and absolute power corrupts absolutely' - 1887.

can swear at them or kick them, with damage only to ourselves. Few of us are exposed to the temptations of absolute power, although our record of cruelty to children and animals could be improved.

Many societies have achieved a high level of culture using slave labour which would have been unobtainable without it. However, the depth of true culture is surely manifested in how such slaves were treated; and, in particular, how easily they could be released from that servitude and whether their children were born slave or free.

2 The myth of inexhaustible supplies

I believe that the myth, or more accurately, the unconscious assumption, that supplies are inexhaustible, has led to certain modes of behaviour in our contemporary society.

There is rampant a kind of insatiable greed. Our standard of living is currently higher than it has ever been for the majority of ordinary people, though probably only historians and the very elderly are fully aware of the massive changes that have occurred over the last hundred years. Far from producing a level of contentment, we seem to want ever more in the way of material goods, diversions, and so on. We want not just a handbag, but lots of handbags, and not just attractive, functional handbags but 'designer' handbags costing £100 or more — an example provided by the 20-something daughter of ordinary working-class parents. She is not atypical, just susceptible to media and peer pressure.

The prospect of immediate satisfaction outweighs any thought of future consequences—'live for the day with no thought for the morrow' is all too common amongst the affluent young. The average current cost of a wedding is now around £20,000 with elaborate 'hen' and 'stag' weekends as part of the ritual. There seems to be a dysjunction between such visible extravagance, and the subsequent reality of finding the price of a decent house.

Because identity, status and self-worth are conferred by possessions, these become more important than relationships; style becomes more important than substance. It is as if *acquiring* has taken precedence over *doing*; passive receiving over the satisfaction of creative struggling.

Of course not everyone behaves in this fashion, but the prevalence of binge-drinking in certain quarters (with its consequential liver damage) tells the same tale. I am mourning the lack of maturity, the failure to achieve the developmental stage we call the 'depressive position'.

The level of debt amongst the population is indeed daunting, and agencies offering support and advice, such as the Citizens' Advice Bureau, are much in demand. Why, in an era of unprecedented wealth for the majority of our citizens, are we such a debt-ridden society?

Our television programmes give us examples of people who have run up enormous debts, not on the basics of living, but on luxury goods — clothes, cars, expensive holidays — often concealing their behaviour from their closest intimates, their spouses or partners. Attempts by financial advisors to help them remedy their dire situation is frequently met by initial massive denial, an absence of guilt and a reluctance to accept the potential consequences of their behaviour. Personal bankruptcies as a way out of such a dilemma are booming.

The times in which we live encourage debt. A government which offers higher education at the expense of a debt burden of many thousands of pounds, thereby mortgaging the future earnings of graduates for years to come, effectively sanctions debt as an acceptable way of life.

The intensive and seductive advertising of credit cards is ubiquitous, as is the bombardment of media advertising of goods or products which promise to transform one's life. The cost of housing is such that people accept ever more heavy mortgages in order to have a home of their own. People feel trapped by circumstances on this issue.

All this makes for very different assumptions from those of the pre-1945 era. Then, for ordinary working-class people, getting into debt was something one did not do because the consequences were likely to be disastrous. I am acutely aware that, under current conditions, I should never have been able to achieve higher education — the prospect of so much debt would have put it beyond the pale. I think there are still people for whom that is true, and the falling proportion of students from social Classes 4 and 5 confirm this. I am angry that it was a socialist government who created this situation.

The roots of financial incompetence

The financial incompetence currently so widespread is reminiscent of the manic behaviour we associate with bi-polar disorder — but in essentially normal people. Is there a common factor? Is it a flight from a threatening depression?

At some deep level, we all have memories of life in the womb when all our needs were met without any effort on our part. If we are fortunate, we have a period after birth when an empathetic mother again meets our needs in response to minimal signalling from the totally dependent small person in her care. Between those two states there is a momentous transition, and how that transition is experienced sets the pattern for life changes yet to come.

I find myself wondering about the birth experience of those who are born by caesarian section — an increasingly common occurrence. It is an essentially passive experience for the infant as compared to the struggle through the natural birth channel. It is a magical transition from one state to another without any active participation on the part of the baby. I think I have detected a kind of passivity in those born by this method, but a proper in-depth study is needed.

However idyllic, or otherwise, our first experience of life after birth, it is still expulsion from Eden. We have to struggle with the reality that we cannot have everything we want immediately we want it. We have to wait; our personal servant on whom we depend is not totally under our control. We don't like it, and we have to develop internal resources for living with this reality. We are asked to do things which do not always come easily; we are offered food about which we are doubtful; we are asked to control our excretory function to suit others' convenience; we become aware of rivals for Mother's attention, whether Father or another sibling. The path of development, of becoming ever more ourselves, is a continual challenge as we struggle to integrate our experiences.

As small beings we need support and encouragement from our carers in these struggles. We need to develop our emotional muscle and experience the satisfactions of new achievements. Others cannot do it for us — but neither can we do it on our own.

My sense is that many children are not receiving the quality of ongoing support — of devoted and consistent parenting — that is needed. The child knows it needs something to actualise its potential, but does not know what it is it needs. It takes whatever is on offer and does its best with that, but what is on offer can be inappropriate — things rather than people, diversions rather than tools.

I see an analogy between this situation and people who compulsively eat too much because the food actually on offer is deficient in vital nutrients. Obesity and ill-health are the result.

The financial equivalent is those who spend, spend, spend on material goods in an attempt to gain what they need from sources that cannot possibly supply it. Passive receptivity has been encouraged rather than engagement in the struggle for an authentic life — because it is more convenient for the carers?

What do the hard-pressed parents do when a massive mortgage demands that both work? How do we escape from the trap we have created for our society? This is not just a rhetorical question. The future health and survival of our culture depends on finding a creative way forward.

Of late much has changed in our world. The international banking crisis of 2008 has revealed that the pervasive lack of reality sense had spread into areas of our social life one would have thought impregnable to such silliness. As a consequence we have been brought up short on our financial roller-coaster, and the myth of inexhaustible resources has been well and truly shattered. We now have the collective task of revising our priorities, and minimising the collateral damage, but a lot of ordinary decent people will suffer in the process as we try to live within our means.

Even now, the phenomenally wealthy USA is debt-ridden to a degree which threatens its standing within the international community, but the rivalries between the dominant political parties has led to a stalemate over potential change.

The threat of climate change has also made us aware that the resources of our planet are not infinite, and that we need to change our mind-set from exploitation to sustainable conservation.

Can we learn?

Our neighbouring small market town has recently experienced a rash of new building. Given the current shortage of homes, this has to be welcomed, the more so since it has taken place mostly on brownfield sites — erstwhile areas of post-industrial dereliction — plus a modicum of acceptable infilling. The style, to my eye, is pleasing and appropriate, mostly terraces and including a variety of styles, giving a higgledy-piggledy, hobbity but altogether neat effect. It is all on a very human scale — what one might call sub-Poundbury.[1]

Not so very long ago it might have been quite other. Wasteful acres of semi-detached villas were still beloved of the speculative builder, and waves of charmless apartment blocks derived from the International style were still being thrown up, even as others were being dynamited into rubble as unfit for human habitation.

The history of the International style is interesting. I could see its appeal following the ornamental excesses of the 19th century, though it always felt more appropriate to a sunnier climate than ours. Flat roofs do not fare well with a high rainfall, and in the British Isles we seem to need to create cosiness in order to survive our winters.

However, the International style originated from the Bauhaus tradition in Germany with a climate more extreme than ours, so its imperatives seem to be in opposition rather than sympathy with its context, its environment. Perhaps herein lies its motive force.

It originated in the years that saw the rise of Fascism in Europe. It was rejected and persecuted by the Nazis as they rose to power, and many of its creative designers fled, to England and the U.S. Nonetheless, there is, at the same time, something almost fascistic about its insistence on people adapting to the machine, the structure, rather than the other way round. The prohibition on decorative detail,

---

[1] Poundbury is adjacent to Dorchester in Dorset, and is famous internationally as a pioneering example of an urban development which challenges post-war trends in town planning.

It was built on land owned by the Duchy of Cornwall, and Prince Charles was instrumental in its design.

on surface clutter, the lack of cosiness, is not how most people choose to live within their homes, and no amount of propaganda does much to change that behaviour. Our style leaders may exhort us, and indeed continually do so, but with limited success — and that mostly amongst the child-less!

Fascism was only one of the authoritarian regimes which came to power in the 20th century. Communism, which purported to be as much on the political Left as Fascism was on the Right, was equally authoritarian. Individual freedom, family ties, personal loyalties, were all subservient to the Fuhrer, the Party, to the ideals, to the imagined future. No sacrifice was too great. Orlando Figes (2007. *The Whisperers: Private Life in Stalin's Russia.*) has documented in dreadful detail how in Russia, even after terrible sufferings, people clung to the dream. Similar stories have emerged from Communist China.

What was it about that time which led to such social aberrations? Europe had been through a lengthy period of industrialisation. Until recent times, with the advent of the micro-electronic revolution and robotic controls, there was no other way that complex machines could operate other than through supervision by people. In the name of economic progress, the lot of the urban working folk was to be subordinated to the machine. At the same time, the mechanisation of agriculture produced a migration of people into the cities, and the consequent depression of the value of labour.

It did not make for human happiness; there was a sub-stratum of discontent, expressed mostly by those less ground down — the middle-classes with ideals. Meanwhile, the political systems had not kept pace with change. In particular, the erstwhile Hapsburg empire was increasingly rigid and inflexible, as was the Tsarist regime in Russia. The comparison with pre-revolutionary France is striking. While there were those who saw the need for change and adaptation at the political level, such change was slow to develop, and events overtook it.

One solution for societies in this perilous situation, an alternative to revolution, is to go to war, which is what England and France did in 1914. Suddenly life looks less complicated, options are clear, the bad can be projected outwards on to the assigned enemy, and the fractious community (more or less) united. In this situation what is needed, and what is invariably evoked, is a strong leader(s) who the community can follow.

I believe it is innate in the human group to look for a leader — a Big Daddy. (The infantile origins within the family are obvious.) Monarchy is one expression of this. The kind of social organisation we have in the U.K., where authority is divided up between a titular head of state and an executive group, is a sophisticated solution, attained over a long period of time and social struggle. We should treasure it! Even here it is temporarily suspended in times of emergency, as in a war situation, when an overt leader takes the helm.

The emergence of Hitler, of Stalin, of Mao Tse Tung and of various African despots is a sign of a failing society — a society which has regressed to a more primitive mode of functioning. In the U.K., in 1945, we were fortunate in having a political system which could rapidly revert to its previous established order. We were fortunate in having an electorate which had the courage, the tough-mindedness, to turn its back on its war-leader, however ungrateful that seemed at the time. We like our political leaders to have a measure of charisma, but are rightly suspicious of too much — and we rely on the monarchy to fill that niche which, for all its human failings, it does rather well!

If the legacy of our current heir to the throne is some sub-Poundbury architecture, the rescue from dereliction of some irreplaceable heritage buildings, the marketing of some first-rate quality comestibles, and the financial support of penniless but enterprising young people via the Prince's Fund, I believe the monarchy is showing a responsiveness both to the entrepreneurial temperament of the English, but also to some basic needs of the community. It is exhibiting a degree of flexibility and pragmatism which is encouraging.

3 The elephant in the room

What no one is saying, yet what is surely at the root of our many troubles, is that the human race is just too numerous. There are too many of us for the planet to support at the level to which we all aspire.

In the attempt to achieve this desired lifestyle we are plundering our limited mineral resources, wrecking our eco-system by exploitive farming methods, wiping out bio-diversity and poisoning the very atmosphere we depend upon for life.

Aggression, war, is continually breaking out in one region or another. As soon as it is suppressed in one part of the globe, it breaks out in another. It is what happens in all animal societies when over-population occurs.

We are developing problems of disease — some new such as AIDS, some hitherto more-or-less under control but now on the increase, like tuberculosis.

We are failing to nurture our young, thereby producing increasing numbers of alienated and delinquent personalities. We are failing to train them in the skills that they need for survival, and that their communities need in order to flourish.

We fail to support those doing the basic work of society, and reward disproportionately too many narcissistic drones with minimal talent.

There has been a massive surge of migration in all directions, but inexorably from the poorest areas of the world to the richest, with the impoverishment of cultures and the creation of often disaffected, rootless people who do not truly belong anywhere. We are inviting disaster.

The ongoing crisis in Gaza is a direct result of too many people trying to live in a limited space — exacerbated by the two different cultures of those peoples. Arguably the state of Israel should never have been created on land which was already occupied by the Muslim Palestinians. What happened was that Israel was established amid ethnic cleansing of the previously settled inhabitants. No one likes to say that publicly, but it is deeply felt by those who were dispossessed. If their bitterness leads them to throw rockets, is it surprising?

Israel came into being on the wave of international sympathy following the terrible fate of European Jews in the Nazi Holocaust. The awful sadness is that their descendants are now perpetrating comparable evils upon others. Some contemporary Israelis, like Hitler's Nazis, are treating other human beings — Arabs — without respect, seeing them as less than human.

The population problem

What next? If we take no action to reduce the world's population, it will be done for us — by epidemics, by famine, by war. We have already seen a serious world shortage of food in 2008 following poor harvests. We live with the threat of global pandemics of 'flu' or bird 'flu'. AIDS is wreaking havoc on some African populations. Communal warfare, such as is currently active in the Middle East. has also been happening in Sri Lanka, and recently happened in Indonesia. Always, terrible casualties are inflicted on non-combatant populations. World-wide sympathy attempts to pick up the pieces; it does not tackle the under-lying problem.

There have been some debatable approaches to resolving the population growth issue. Pol Pot initiated an attempt which involved murdering a substantial proportion of Cambodia's people — one kind of 'green' solution!

The Chinese government has had a ban on more than one child per family. We can only guess what this means for the parents. Recently a woman was accused of attempting to murder her one child and her step-child in order to have a another baby by a new husband. An extreme case, but perhaps the tip of the iceberg of frustration. We do not know how this will impact on the dominant personality type produced by the culture: only children, singletons, do have their own distinct characteristics.

There is much evidence to suggest that the surest way to reduce the birthrate in a society is to educate the women.[1] If young women are given other interests, aspirations, ambitions, their energies are diverted from overt sexuality and child-rearing. They tend to postpone having their babies until later, and have fewer or none. This is already a marked trend among the middle classes of the developed world, where the average age of a first-time mother is such that they would have been labelled 'elderly primips' two generations past.

This has its downside. Caesarian section, replacing normal labour, is widespread. There is a considerable problem of infertility. Older mothers may have more wisdom but they certainly have less energy, and are unlikely to be available as hands-on grandparents in their turn.

Are women who have had interesting lives outside the home willing and able to deeply engage with their babies with the total involvement that is needed, albeit briefly? Mechanical or substitute child-care is a poor exchange, and impoverishes the developing young life. As a society,

---

1 Is this why the Taliban is executing those who try to teach their girls? Given the rate of population attrition in the chronically warring society of Afghanistan, perhaps different priorities are to the fore there.

we have not yet learned to value the maternal role as it deserves.

A recent report emanating from The Children's Society (published 1. 2. 2009) makes chilling reading. It says what many of us have felt for some time but have been unable to admit to publicly, namely that the selfish culture of personal advancement, of material acquisition, of working mothers and long working hours for both parents, has short-changed our children. They are said to be the most unhappy children within Europe, and ill-mannered and undisciplined to boot.

One would like to think that, as our birth-rate declined, children and child-rearing would become more valued, more worthy of the time and care of intelligent people, not just trophy acquisitions, not chores to be fitted in amongst other more important affairs.

Liberating women from total absorption in rearing the next generation has serious consequences for the menfolk. Given the opportunity, women compete with considerable success. When educated together, girls tend to outshine the boys for much of their school career. They are emotionally more mature for a given age and tend to be more fluent in talking and writing. Our predominantly literary educational system favours just these characteristics.

Girls are making ever-greater inroads into the professions, and their capacity to multi-task give them an edge in many fields. There is a real danger that the medical profession may become women-dominated as has the teaching profession, to its detriment.

The crisis for men

Where does this leave the menfolk? I fear a backlash. I sense that decent men are struggling to come to terms with these trends, but that does not mean they are happy with them. There is an under-current of anger, not helped by the excesses of the early feminists (now happily fading), nor by the constant depiction on television adverts of the menfolk as amiable fools.

Men have greater physical strength than women, but this is needed so much less in our kind of society where machines have replaced brute labour. They tend to be better at mathematics, and to have more developed spatial imagination. To date, this has contributed to their dominance in such fields as engineering, where these attributes are core skills. They are generally more aggressive, and in the past the ability to defend the tribe's territory and wage war was a vital attribute. However, these tendencies are as much a liability as an asset in today's world. In the past they have proved an invaluable resource, but in a world where warfare is going to be waged largely through military robots, they will be history.

Recent military skirmishes have demonstrated the limitations of personnel on the ground. When used, the skills required are more those of policemen, rather than the soldiers of yesteryear. And women can pilot helicopter gunships as competently as men, as was noted in the first Gulf War.

Men are frightened of women — something too little understood. They are frightened because of their need for them. It makes them vulnerable. They were frightened of their dependence on their mothers, because that powerful bond had to be broken if they were ever to become real men.

They are frightened of their partners because of both their need of them and their need to protect them. It undermines their courage to battle the enemy, and undermines the male bonding which is vital in adventure and war. They are frightened too of the woman's fertility

— of its primitive power, of something beyond their control.

The structures of a patriarchal society allow these fears to be managed. While the women were totally absorbed in their biological role, some kind of balance was maintained. If the women lived restricted lives, there were often secondary gains to their state. That balance has gone. Women have escaped. If their fertility is to become less and less needed, what then? What does that do to the relationship between the sexes? How will each see the other? What will each need of the other? It cannot be quite the same.

We are in unknown territory. We need to think carefully and long. We need to respect and value basic biology because it will not be ignored or over-ridden. It will exact its demands, if not its revenge. Above all, we need to respect each other and those innate sexual differences, differences of function and mind-set, which at their best are complementary, creative and give mutual support.

4 Depression in men

On January 9th, 2011 an extended article in *The Times* reminded us that, although less common than in women, depression is nonetheless a widespread affliction amongst men, and the major killer of young active men[1]. It may go unrecognised because men are less willing to admit to it, to talk about it, and to ask for help. Their sense of shame and inadequacy all too readily paralyses any remedial action they might take.

The first-hand accounts of personal journeys made painful reading as a number of men, who had achieved much in their lives, described their struggles against a debilitating undertow of recurrent depression.

Amongst them, the jury remained out as to whether it was psychogenic in origin, or a physiological imbalance and therefore a 'real illness'. They appeared to have coped in the end by accepting that it was beyond their control, to use whatever medication that was on offer, and to cling on to the intellectual knowledge that the bouts did not last forever.

While none of the men interviewed fell into the category of the more severe illness that we now call bi-polar disorder, I found myself thinking of that condition and its origins.

Aside from Melanie Klein's ground-breaking paper on the infantile manic defence, I have found surprisingly little

---

[1] by suicide

physiological in origin, is meaningless if its origins are very early, before the psyche-soma has properly differentiated. One would like to know something of the quality of mothering which they received as infants. Extreme emotional distress before the age of two years affects the developing autonomic system, and the resulting changes subsequently act as firmware.

A father who is available during the period when the infant is beginning to develop a sense of separateness, (18-36 months) is crucial to his development. He needs that third person to help detach himself from the embracing ambience of mother. If that mother has her own problems — if she herself is depressed, or finds it difficult to allow her child to grow up — the father is even more vitally necessary.

Good fathering begets good fathering. Without it, the men struggle to fulfil a role for which they have had no apprenticeship. No wonder some of them find it tough.

5 The politics of gender

'Why can't a woman be more like a man!" The words are uttered by Professor Henry Higgins in the musical comedy *My Fair Lady*. It wasn't that Higgins hated women. On the contrary. Insofar as he was aware of them as people at all, he liked them, as long as they ministered to his needs like his mother, or slaved for him like Eliza Dolittle, though of course, they never performed their duties as well as he thought they should.

What he could not cope with, in fact feared, was their emotionality. At the end of *My Fair Lady*, Higgins and Eliza pair off. In the original play *Pygmalion* on which the musical was based, that shrewd realist George Bernard Shaw has Eliza marrying a young man of her own age with whom she can, hopefully, have a relationship of some mutuality, and where the power balance is, if anything, weighted in the opposite direction.

Men's fear of women's emotionality is widespread, profound and deeply-rooted. What is at stake is their masculinity, their phallic pride, their autonomy. If maintaining their male identity means demeaning women, that is what they will do. It is often accompanied by a compensatory idealisation of women as goddesses; but the idealisation is two-faced. The Irish goddess Brigid was ferocious; the Hindu Kali had a girdle of skulls. Men's fear of women is an unconscious memory of their total infantile dependence on a woman who was both wonderful and terrible. No wonder the cult of reason calls.

The siren voice of the rational man, the rational life, the rational society is powerfully seductive. Emotions are messy — wonderful, yes, painful frequently, all too often disturbing and chaotic. Oh for the serenity of a life from which they are excluded! No women, no sex, but order and reason and good sense. Beat a retreat to the male club, the cloister, the laboratory, the ivory tower where one can live the calm, rational life.

However, in doing so, one may be protecting oneself from all kinds of difficulties, but one is limiting one's development as a full human being. Moreover, Dionysus denied has a way of liberating the Furies, as the Ancient Greeks knew.[1]

The courage of women

A recent radio item disturbed me.[2] Superficially it was good news. Black girls are doing well in the work-place. They not only outshine their menfolk, but — the point of the item — they, on average, earn slightly more than their white sisters.

It was suggested that black women are strong personalities with a strong work ethic. Unlike their white sisters, they do not expect their menfolk to keep them, even when they have children to rear. They pass on this expectation of self-sufficiency and achievement to their daughters who then follow the same pattern. They do not experience the same barriers of racial prejudice as do their menfolk. Hence they are flourishing. Bully for them!

So why does it bother me? Firstly, without being pedantic, I object to and am confused by this term 'black'. It is applied too loosely to have significant meaning. Were they

---

[1] The reference is to the Ancient Greek play by Euripides' *The Bacchae*. In it King Pentheus, while fascinated by the Bacchanalian rites, denies that Dionysus is a god, and in revenge is torn to pieces by the Furies.

2 Woman's Hour: November 2008

talking about Asian girls? We see many of these young women nowadays as journalists, in the media, on television, looking attractive, competent and composed. While assertive when need be, they impress one as agreeable, on good terms with their femininity, and non-threatening to their male colleagues. One gathers that they are represented in other areas of professional life too, as doctors, lawyers, financial personnel. They have the confident aura that suggests a solid supportive family background and a good education.

I do not think, though, it was with these women that the item was concerned, but rather with those from an Afro-Caribbean background. Here the family pattern is quite different. For reasons that I suspect relate to the devastating historical experience of slavery, the menfolk are all too often not around as part of the family unit. The women have learned to manage without male support. The acquired strengths that have enabled them to survive are paying off in the workplace.[1]

While saluting their courage and stoicism, I cannot feel it is a scenario to celebrate. The dominance of the women has been disastrous for their menfolk. The boys all too often have no adequate role models, cannot see that they have a significant function, and the drift towards delinquency and the drug culture is all too seductive.

Nor is this an English phenomenon, nor even a 'black' one. A recent statistic suggested that of babies born to 'black' women in the USA, 75% were to single mothers. There, as in England, the deleterious effect of the absent father is now well-documented with its links to delinquency in boys and criminality in young men.

---

[1] Emmanuel Todd in his book *The Explanation of Ideology* writes persuasively about the connections between marital relationships and social structure. His observations about the patterns in Africa are relevant to the above comments.

Of course there are many exceptions, where men strive to be good husbands and fathers, and the strong tradition of Pentecostal Christianity in the Afro-Caribbean community is a positive cultural force in supporting families.

There are, too, sub-cultures where matriarchy is a settled pattern which works for all parties. The Welsh *Mam* is a force to be reckoned with, as also is the traditional Jewish mother. However, in both these cases, the gender roles are well-defined and the menfolk have their own significant part to play. In Wales, if the women are assertive and managing, it is the father who tends to show the tenderness, the empathy. He has his own honoured role as breadwinner, as rugby player, as poet. Similarly, the traditional Jewish father too has his own areas of cultural life well demarcated and valued.

Both groups give great weight to their joint effort of child-rearing, and it shows in the disproportionately high levels of cultural achievement of these numerically small groups. However, is there perhaps a link between the oppression which these two groups have historically suffered, and the strength of the womenfolk which has developed?

I have been delighted to watch the emergence of women out of the shadows of the 19th century paternalistic society. Our lives are so much fuller and freer than those of our immediate forebears if they came from any but the wealthiest sections of society. But... we have paid a price. Our menfolk are paying a price. We need to look up and take note of what is happening in our society, and do a little critical thinking.

Kept women?

I do not know of many white women who expect their menfolk to keep them. Most of the younger women I know are all too aware of the fragility of contemporary marriage to be complacent. But some of them still understand that good parenting and good child-rearing demands time, patience, love and, if only temporarily, some sacrifice of personal achievement. For all our cleverness, we cannot be in more than one place at once, and we cannot serve both God and Mammon. Hard choices have to be made.

Someone has to do the mothering. Someone has to create and maintain the supportive facilitating environment. Someone has to hold it all together, remember the nitty-gritty detail of daily life, fill in the gaps, be there during illness and crisis. That mindset does not sit well with most of the work available out there, nor is there the necessary flexibility on offer from most employers.

Grandparents can be a godsend, and where extended families live geographically close together, this can work well. However, the home-based grandmother is not the reliable phenomenon she once was. Nowadays, she too may be working.

Behind all this is what I see as our cultural blindness to what is traditionally women's work. It has too long been taken for granted, unseen, unrecognised, unpaid, undervalued — a part of the wallpaper, so to speak. No wonder many women prefer to put their energies into activities which bring them money, status and recognition, a sense of identity and a measure of independence.

However we choose to order our lives, however as a society we choose to reward certain activities, and undervalue others, there is a price to be paid. We should ask ourselves what that price is.

6 The crisis of authority

I was on one of my weekly trips to London, there on professional business. During a quiet lull in the afternoon I was walking in an elegant part of the metropolis, a well-manicured garden square protected by railings on my left, a terrace of handsome white stuccoed houses on my right. Suddenly a young man ran past me on the opposite side of the road, pursued by two other youths of similar age. From their uniform dress, they were senior students at a nearby school.

The two pursuers caught up with the first and began to attack him with their fists. It didn't feel friendly; it felt ugly, vicious, but there was no noise, no complaint, no shouting for help. Should I intervene? They were well past me by this time and there was no one else around. If I had turned back, I doubted if my participation would have been welcomed by any of the parties, and there was also the outside chance I might have been attacked myself. I moved on feeling suddenly vulnerable, uncomfortable and vaguely guilty.

When I returned home, I mentioned the incident to my son-in-law who pooh-poohed my scruples, dismissing the incident as just the horse-play typical of that age group. I was only partially reassured.

That incident took place more than twenty-five years ago. Since then, there has been a distressing spate of teenage killings, commonly involving knives or guns, in the poorer areas of our cities. However, that memory inclines me to feel that the more recent appalling violence has been brewing for a long time. It also makes me wonder at the reasons, excuses, that are commonly offered as explanation — poverty, deprivation, broken families, ethnic sub-culture. Maybe.

The origins of violence

I think there are other issues at work in our culture which are surfacing in this form — that teenage violence is only the presenting symptom of a deeper malaise.

There is some evidence now that children who are cared for in day-nurseries from a very early age develop greater social skills than those more isolated from their peers, but also that they tend to be more self-assertive, even aggressive, in their responses. (Hardly surprising perhaps. How else do you get noticed in a crowd?)

There is also a lot of evidence that small boys who have no active father figure around as three to four-year olds tend to become aggressive and unmanageable. At this age, they are testing the boundaries and need to be contained, and it is not easy for the mother to be both nurturer and container. In an age where fathers often work long hours, and there are many separated parents, this scenario is not uncommon.

While school can provide a firm containment, teachers are nowadays restricted in the limits they can legally impose; they no longer have much in the way of tools to manage the unruly and disruptive. Also, education has increasingly come to be in the hands of women. There are few male teachers around, particularly at primary level, and the whole ethos has been slewed towards a feminine, nurturing approach. While this has been a welcome corrective to the often harsh discipline of previous generations, it is not always enough. A firm male hand has its place.

The current educational style seems to favour girls rather than boys, and where co-education is the norm, the girls are achieving more in the examination system than the boys. Add to this the basic biological fact that girls for many years are age-for-age developmentally ahead of boys, and the boys' situation begins to look daunting.

As girls' self-confidence grows, as their aspirations push them towards independence, both financially and emotionally, where does that leave the young men? I suspect many of them are left wondering what function they have in life.

In some of our ethnic sub-cultures, we are seeing a distressing spate of so-called 'honour killings', as the young women absorb the attitudes and expectations of the host culture. Their menfolk find this intolerable; and while the host culture does not sanction a murderous reaction, I suspect that the underlying rage is not so very different amongst ourselves. It just finds a different outlet.

Boys for the most part are naturally more competitive than girls, their narcissistic male pride makes them vulnerable to loss of face, and the testosterone surge of puberty enhances the challenges facing them.

Young men have always been unruly, with a propensity to aggression — nothing new about that. (Read your Shakespeare!) The habit of aggressive acting-out is also spreading to young women in some parts of our society. Other cultures have mobilised the male aggression in various ways, by creating hunting parties, fighting bands, going to war. The last option is increasingly unacceptable and unavailable with the irresistable growth in lethality of our military technology.

Containing aggression

Our dilemma is two-fold. Firstly, we are in denial about the various factors in our society, as outlined above, which are currently tending to fuel the problem. Secondly, we seem unable to find, and exert, the authority needed to contain the aggressive tendencies of our young.

I believe that we have a widespread crisis of authority in our society at all levels. It makes it increasingly difficult to recruit people to take on leadership roles, and when we find them they are continually undermined, particularly by our destructive press and broadcast media. The present crisis in the recruitment of head-teachers is one such.

Also, the romantic cult of the rebel has a lot to answer for, and I believe that its fashion has now outlived its usefulness. We are in dire need of some positive, paternal heroes as opposed to stroppy adolescent ones, though that too brings its dangers.

The grandiosity evoked by successful empire building was still sufficiently active in England at the end of the nineteenth century for Germans to comment unfavourably on it. However, by this time, Bismarck had transformed from relative poverty not only the Prussian educational and industrial systems, but also the Germans' perception of themselves and their place among nations. This changing balance of European power foreshadowed the clash of 1914-18.

In the event, Germany was defeated, but as in all wars, everyone lost. Great Britain lost a sense of security and inviability. It also lost a vast swathe of young men of all social groups, but disproportionately from the officer class. A generation of young women were destined to remain unmarried, and needing to support themselves financially. A generation of children grew up without fathers, and with mothers who had to struggle with grief, financial hardship and loneliness while trying to do their best for their young. A generation of small boys grew up without fathers as role-models of what it was to be a man. Photographs, medals, stories, filled in some of the gaps, but they were no substitute for the daily living presence. Mothers clung, understandably, to their sons and found it difficult to let them go so that they could find their own mates. Sons felt responsible for their mothers and sisters, and caught between sometimes conflicting demands. Young wives could easily feel resentful at the intrusive presence and demands of the older women.

It was nobody's fault; but it did undermine something vital in family and social structure. It undermined the young man's ability to assume proper authority and initiative. There was no one to show him how to do it. There was no father to subtly help dissolve the powerful mother/son relationship. There was no one to encourage the mother to let go of her child and her power. Why should she — there were no compensations waiting for her! There was often no one to help the young man find a role in the wider world, and assume his place there.

Of course, some men managed to surmount these hurdles, but many didn't, and it left their wives vaguely dissatisfied with them as husbands. The developments in women's education, the increase in their earning capacity, and then the fashion of Women's Liberation added to the tendency to denigrate the menfolk.

The men have responded by withdrawing their commitment, limiting their emotional involvement, substituting casual sexual relationships for anything more substantial which might be too challenging. New generations of abandoned, fatherless children are being created. Those fortunate children with stable families become the envy of those without them. Sometimes they then become the victims of the underlying, apparently incomprehensible, rage. Once again the best of the generation is slaughtered.

Am I attributing too much to the aftermath of war? I don't think so. We have suffered two costly wars within just over thirty years — a generation. The appalling squander of men in 1914-18 was the worst, but the losses of 1939-45 added to the damage. Even where the soldiers returned, the pattern of family life was often irrevocably changed.

We cannot go back, only onwards. How do we facilitate the long-term healing of our damaged society? The fun of the Flapper era, post-1918, and that of the Swinging Sixties and Seventies, was in my judgment a social form of manic denial. Casual sex, recreational drugs, mind-numbing music, can be a substitute for real feeling. If the pain is too great, we cannot face it. It is still, for those of us who lived through that time, almost unbearable to look at the reality of the Holocaust, the excesses of Stalin's Russia. It freezes emotion, rationality, creativity.

Real social change is slow — 'unto the third and fourth generation.' It will be the young ones who will have to find their way through the social chaos and rubble. If we can help settle the dust which impedes their vision and chokes the breath, that may be the best that we ourselves can do.

Practical politics

Our political system favours the open-loop approach — the decision maker, the charismatic leader, 'the man on the white horse,' the hero.[1] He issues his commands but has little interest in receiving feedback from those he rules, except at election time.

Open-loop control is a very masculine way of 'doing' — the Man on the White Horse is surely the Father who traditionally swoops in from time to time to sort out

---

[1] Would this factor be less seductive if we had proportional representation? We would need to establish a minimum qualifying level of 10% of vote in order to keep out the lunatic fringe, and to avoid tiny extremist groups acquiring disproportionate power as has occurred, for example, in Israel.

problems. Women's mode is much more that of closed loop control, modifying what they do in the light of the perceived consequences of their actions. They use it all the time within the family context.[1]

As a society, we have for a long time been neglecting our infra-structure in many areas of life. In the crucial area of the development of our citizens, we have:-

- failed to train enough midwives, and paediatric nurses.
- allowed the health visitor service to decline to near-nothing.
- encouraged mothers of young children to work.
- failed to give them any sense of the vital nature of their parental role, or the respect it deserves.
- failed to ensure adequate affordable housing.
- failed to reward the commitment of marriage.
- failed to protect the sensitive developmental period of 'childhood' by allowing excess exposure to commercial pressures.

We have been denying biological reality in favour of the fantasy worlds of celebrity and consumerism. We are now paying the price. In the summer of 2011 many were astonished to find the streets of our cities out of our control for a time, a prey to riot, looting and arson. Why the surprise? Surely we had it coming.

---

[1] It would be interesting to look at women-only businesses, such as Virago, the publishers. Do they operate differently from others? It sounds like a research project for someone!

7 Post-imperial blues

In the fireplace of my living room sits a *Jotl* wood burning stove. Matt black and elegant, it can be seen either as austerely contemporary or alternatively as reminiscent of a mediaeval knight's helmet. It is Norwegian in origin, simple in design and ridiculously easy to manage. In the winter season it becomes the warm heart of the house. I am told that its original prototype was made during the Nazi occupation of Norway, when fuel was scarce but the forests were an available resource.

A recent visitor to Norway was telling about her holiday there, and describing the capital city of Oslo. Built overlooking the fjord, it sounds a delightful situation. Apparently the inevitable urban transport problem has been solved by driving a tunnel underneath the city from one side to the other. She also commented on the number of open green spaces, and the striking sculptures to be found within them.

My knowledge of Norway is limited, but the few Norwegians I have met have been cheerful, positive people, temperamentally more like the Danes than the rather dour depressive Swedes.

Then, we tend to forget that Sweden once had a great empire extending into Russia and down to the Black Sea. For centuries they were the scourge of Northern Europe. It must have come hard when that power waned.

Building empire is a dubious activity. The expansionist phase produces a manic high in the conquerors, but the subsequent collapse is devastating. It leaves behind not only the depression of defeat, but an infra-structure distorted by military demands, attitudes and expectations out of tune with reality, plus the relative poverty which follows upon squandered resources.

The Arabs have never recovered from their loss of empire, never achieved the heights of civilised living, the sophisticated science and medicine, the artistic flowering, of their time of greatness. Some still hanker for a worldwide 'umma'.

France still yearns for *'la gloire'* of Napoleon's time, seeing him as a great national hero, ignoring the reality that it was he who created the *'levée en masse'* which led to the waste of so many young lives, in his own time and beyond.

The USA is beginning to count the cost of its military adventures, and its subsequent besmirched reputation amongst nations. The Guantanamo Bay prison will be long remembered, as will the inadequate response to the hurricane Katrina which devastated New Orleans in 2005. Rhetoric about liberty, freedom and democracy has a sour ring in the light of recent history.

Britain, post-empire, was left with an educational system designed to produce an active military hierarchy and a colonial civil service. The ruling classes had ignored a major source of the wealth needed for its enterprises — the innovation that fuelled the industrial revolution — since that had been created by Nonconformists and other outsiders excluded from the university system.

The teaching of science and technology is still suffering from the low esteem and chronic under-funding secondary to a colonial perspective. So, also, are the Arts, squeezed out by the profligacy of a war culture. Wealth has been poured into military dreams, while our infrastructure decays.

Farmers have been squeezed off the land at an alarming rate, some into suicide, their skills and experience lost. A whole generation of young professionals cannot afford to get on the housing ladder and continue to live in shared accommodation like students. They opt for late pregnancies, run into problems of infertility, cannot themselves afford to care for their infants during those crucial early formative years. There is a chronic shortage of nurses and midwives, and the NHS is kept going by importing skills from abroad.

Our metropolis is currently serviced by people many of whom cannot afford to live there, but travel many miles to reach their place of work. Now, it seems not even the fighting soldiers are adequately equipped, housed or subsequently cared for.

Among the great nations are a handful of smaller ones who have kept their heads down, have avoided war wherever possible, used such natural resources as they have to good effect, fed their minimal wealth back into the community, looked after their citizens. The names that come to mind are Norway, Denmark, Finland, Switzerland, Portugal, New Zealand. When to fight has been unavoidable, they have made their contribution, but have not sought out conflict. The result is that they are surely amongst the more agreeable societies in which to live.

Is some subliminal recognition of this is behind the current trend towards devolution within the British Isles? From the eighteenth century Great Britain — the United Kingdom — was always a political creation. Many English people are surprised and unbelieving when they discover that the Welsh have never ceased to feel themselves an occupied nation, yet have maintained a separate cultural identity against all the odds. Scotland was tricked into the Union by the Darien fiasco at a time when England needed to present a strong front against the French.

Both Wales and Scotland can now imagine an independent future for themselves within a united Europe. Scotland has enormous potential of natural wind and water power if developed wisely. Wales has as much (if not more) natural wealth as Norway, with a comparable population.

England is still struggling with the aftermath of the First (1914-18) and Second (1939-45) World Wars, and the recent military adventures in Iraq and Afghanistan have only added to its burdens. The well-intended but too massive influx of immigrants has compromised its social cohesion. The Celtic fringe has attracted fewer recent migrants but, with a long tradition of sea-faring incomers to draw on, has integrated better those it received.

Smaller societies can operate more at face-to-face levels. Leadership becomes more accessible and more accountable. The active participation of all adult citizens becomes more possible. Democracy then has meaning. By contrast, when military adventures are the priority, there has to be control from the centre, there has to be a dominant bureaucracy. The parliament at Westminster is still operating in this mode, and people are tired of it. It is acceptable in a crisis; it is no model for a civilised peace.

The ideals of socialism are admirable. The practice has not worked because the basic unconscious assumptions have been faulty. A gaping ideological void has been left by its death. What now for the Left?

8 Socialism

It is painful to acknowledge the failure of Socialism. For a generation who saw it as the answer to men's dreams of a good life, of equality, of the end of poverty; who remembered their history and that of their forebears, the suffering, the deprivation, the callous uncaring of those who had for those who had not; it is well-nigh impossible to let go that dream of something better.

But the truth is that it has failed. In its many manifestations, it has failed. Whether the kindly benevolence of the post-war Attlee government, the monolithic oligarchy of Soviet Russia, the monstrous regime of Pol Pot, the benign bureaucracy of Sweden — all have created, not heaven on earth, but an unbearable mess.

It has failed, not because its principles are mistaken — as aspirations these still shine bright and clear. It has failed because it has operated from underlying assumptions which are just plain wrong.

Behind it are the philosophical notions, propounded by the philosopher John Locke (1632-1704) that human beings are tabula rasa — that they can become, or be made into, anything we choose. The idea of the New Soviet Man typifies this thinking. In a more benign form, the Gestalt theory of the New Age hippies was experienced as individually liberating until too many of them ended up in middle age as poor and lonely.

Human beings are mammals, social animals, with a very long period of maturation. We are born with certain innate characteristics, needs and potential. How these are developed depends on the facilitating environment in which we are born and reared. While these innate characteristics can be modified, they cannot be changed, only lived with. They will not be denied, and attempts to do so are disastrous. The insights of Euripides "Bacchae" were powerfully to the point.

Socialism has flourished as an idea during a phase of social perversion. In so many areas we have witnessed a denial of reality, of difference, of limits and limitations, of traditional boundaries and hierarchy. Individually it has often been fun. Collectively it has been disastrous.

- We have the highest rate of teenage pregnancy in Europe.
- We have generations of families who have never worked.
- We have too many men who are sexually active but renege on those commitments which make for maturity.
- We have an inability to assume authority, and a failure to exercise it with wisdom and discrimination.
- We have infantilised our population, and opened the floodgates to the current massive intrusion of bureaucracy into our lives.

We shall suffer the fall-out from this phase. We shall not prevent a repeat or devise a better social organisation until we have greater understanding of how human beings actually tick — as individuals, as small groups, as large groups.

Politics mobilises very primitive social forces. Politicians succeed to the degree that they are in touch with, and can manipulate, the zeitgeist at a very primitive basic level. Sooner or later they fail because their life-style makes it inevitable that they will lose touch with that zeitgeist. This is why they are potentially dangerous. It justifies our democratic system which allows us to get rid of them at frequent intervals. However, we still need leaders — we need father figures, patriarchs of the tribe. It is deeply engrained within us. It would be less damaging if we were more conscious of it.

It is the unconscious basic assumptions which lead us astray. It is the denial of reality, of basic biology which we need to address. On that front, the feminists have done us no service. Happily the younger women of our human tribe are beginning to think again about their priorities. Thank God for biology!

May 2010 was the point at which the thirteen-year government of New Labour came to an end, and was replaced by a coalition of the Tories, led by David Cameron, and the Liberal Democrats, led by Nick Clegg. For most of that period, the Prime Minister had been Tony Blair, who began his premiership with an enormous parliamentary majority and much goodwill.. By the time he gave up the office to his long-time colleague Gordon Brown there was a sense of disillusion in the country, the Labour Party appeared deeply tired, and the final decline was swift. For Gordon Brown, an intelligent but flawed man, it quickly became a personal tragedy. New Labour had gone. But it needed to happen.

Bureaucratic centralism

At the heart of socialism is the desire to control everything from the centre. How else is radical change to be made? It doesn't work. It cannot work. The systems are too complex to be micro-managed, and the attempt to do so leads to progressive failure, the suffocation of initiative, and the infantilisation of the populace.

A defining characteristic of New Labour was bureaucratic centralism, the urge to rule from above downwards. The English do not like it, and in 2010 sent the Labour Party into the wilderness. having reached the point where they no longer found it tolerable. In the long perspective of English history, this belief in bureaucratic centralism stems from the Norman invaders, who practised it ruthlessly on the indigenous population. It was a form of social organisation quite alien to the Saxons who had a system of common law which allowed for local variation and interpretation.

It was equally alien to the Celtic remnants who were later subdued and integrated into Greater Britain. Both Welsh and Scots had a form of communalism which can look superficially like socialism (and historically has joined forces with it) but which has quite different roots — in the clan system that was their heritage.

How ironic then that it is the Labour Party — the party of the common man so-called — which has inherited the mantle of the Norman conquerors, while the contemporary Tory Party now stands for the pragmatism and de-centralisation which the subjugated Saxons would have approved.

Both the Welsh and the Scots, having been granted a measure of devolution, are both moving steadily in the direction of independence — not quite what was envisaged by the UK government at the time devolution passed into law. Having been seen as the bulwark of socialism in these islands, and hence the essential mainstay of the Labour Party, they are now drifting away to form other alliances — the Scots with the Nationalists, the Welsh with Plaid Cymry. It is difficult to see how the Labour Party can remain the strong political force it has been ever since the decline of the old Liberal Party. Even its working class roots in England have shrunk with the decline in numbers of the traditional manual working class.

The basic underlying assumption of Socialism is Locke's belief — that people are born as tabula rasa, and can be made into any shape by appropriate external circumstances. It also drove the 18th century French Revolution, the subsequent American break with its colonial overlord, and in the 20th century the Russian upheaval. An inadequate model, it has long had its day.

We now understand people as being formed by the subtle and complex interaction between their innate endowment and the circumstances of their lives. We are all different. We each have inborn potential and talents; we each have a specific life-experience; we each have something unique to contribute.

The role of the nation, and therefore of government, is to facilitate as best it can the potential of its citizens, recognising that ultimately its people are the only wealth it has. This requires as much liberty as possible within a framework of law. It requires a recognition that the earliest years of life are formative to a degree impossible to exaggerate, and that subsequent interventions are costly and relatively inefficient. It requires a radical re-think of perspective and of priorities.

Another underlying assumption of Socialism — that of equality — is obvious nonsense unless interpreted in a poetic or religious sense. In politico-social terms, it is a defence against envy.

The rumpus over MP's expenses exposed irregularities among members on all sides of the House, Tories quite as much as others. However, it was the ruling New Labour which suffered most in public esteem. The level of bitterness revealed in their former adherents is perhaps understandable, but it does expose an unlovely degree of envy of the 'fat cats'. One cannot help feeling that it was the fear of this envy which has inhibited various governments, of both persuasions, from rectifying the unrealistically low salary scale of MPs. As so often in life, they have brought about their own destruction!

Socialism was strong on diagnosis. Its insights into the ills of our society remain valid and challenging. Even so the treatment it proposed has failed, not once but repeatedly. Bureaucratic centralism is in conflict with biology; another way has to be found.

9 Democracy

Democracy is an achievement of a mature society which has struggled and suffered for a long time in order to create it.

In the UK we can see its beginnings in the confrontation of the barons with King John which led to Magna Carta. We can follow it down through the Civil War which lost Charles the First his head, and the subsequent turmoil of the Protectorate and Restoration; through the episodes of the Tolpuddle Martyrs and the Peterloo Massacre; through the battle for universal male suffrage, and then for women's suffrage less than a century ago. It has needed a deep, ingrained feeling for justice, a sense of the value of each and every man, and the courage to personally confront tyranny and risk the consequences. It has taken us a long time to get there.

It also requires enough maturity in the populace to sustain it — an acknowledge of the value and needs of the other, a willingness to restrain one's own demands for the common good, a capacity to compromise when needed, a recognition that no man is an island.

A brief glance at those countries which have a steady tradition of democracy will tell a similar story. They have been a long time getting there. The shortest gestation period of any is that of the USA which drew heavily on the experience of France, and recent events leave one wondering how secure it is there.

The need for autonomy

Democracy cannot be imposed from outside. It has to come from within a society when that society is ready to pay the price. It cannot reliably be learned from anyone else. Structures can be copied, procedures can be put in place, but if the people themselves are not ready for it, it simply will not hold.

We have seen this in post-colonial Africa. We also watch the current struggles within the young Pakistan, and compare them to those of their neighbour India, with its immensely long and turbulent history. We mourn the fragility of democracy elsewhere as, all too often, autocrats seize the helm of power. We should not be surprised.

What is distressing are the current attempts to carry democracy to others by armed force., as in the occupation of Iraq by American and British forces, following the toppling of Saddam Hussein's regime.[1] Disorder there was and is still rife. Imposed democracy is a nonsense. It cannot be done,

To believe that it can shows a lack of maturity in those who attempt it. They are like adolescents — full of idealism and good intentions, certain of their own vision and the wrongness of others, believing in their own power to remake the world.

Sadly, too much of our world diplomacy has been in the hands of individuals who have all the charm, and all the limitations, of adolescents. It is time we recognised what we are looking at, and asked ourselves why we have fallen for their enthusiastic rhetoric. Even worse, in that process we have willingly colluded with an erosion of some of those very rights and limitations which are at the heart of democracy.

---

1 The establishment, by the USA's Bush administration, of the Guantanamo Bay prison camp was a source of shame and dismay.

Open loop, closed loop

Our political system favours the open-loop approach — the decision maker, the charismatic leader, 'the man on the white horse,' the hero. When these leaders fail, as they invariably do, the current situation of a major election every five years produces a much too long delay in the feedback loop, and results in the kind of 'boom and bust' or 'hunting' mode we have experienced many times in the economic field.

A re-election of 1/5th of the M.P.s every year would make for the minimal changes comparable to what the Bank of England now does for the economy. Also, it might encourage M.P.'s to keep talking to their constituents! It might also be easier to control the disgracefully large amounts of money spent on elections, and the muddled and dubious practices used to raise that money.

We should beware! Our political institutions show only slight indications of functioning more effectively as yet. Four years on from the invasion of Iraq and we are mired down yet again, politically and militarily — in Afghanistan and Libya.

However, perhaps something has been learned in the meantime. To date, there has been a firm refusal by William Hague[1] to take any action over Syria in spite of its governing authorities' heavy-handed response to protest, with many civilians killed.

While attempting to save face, we are pulling out of Afghanistan as fast as we can with the message that it is now up to the locals to solve their political problems.

In Libya, our commitment to intervention has been less wholehearted with lower expectations, and the unwillingness of some members of NATO to participate at all. We have had to bear with the slow and chaotic

---

[1] Our UK Foreign Secretary at time of writing.

attempts of the local populace to overturn the old regime. We have had to listen to the news of fearful atrocities as they have unfolded. Whatever the final outcome — and at the time of writing it looks positive — any achievement will be essentially that of the people themselves. The strife and suffering they have undergone will have honed them as a society; it will have left its mark.

The loss of life amongst our soldiers in Afghanistan, and the relentless media coverage of the most terrible international events makes it painful to accept that there are some situations that cannot be solved by outsiders, however well-intentioned.

10 Social identity

In far pre-history, our identity was as a member of the extended family hunter-gatherer group. When the human race took up farming, we added a location to our sense of who we were; we belonged in a certain place. Although there were always those who travelled and, as we now realise, travelled extensively, whether as traders or Druids or missionaries, they acquired another kind of identity by virtue of their occupation. Urbanisation and multi-cultural mingling was brought to these shores by the Romans, but the phenomenon did not outlast the end of empire. It took another thousand years, and the collapse of feudalism before widespread conurbations began to develop again. Even then, most people still lived a restricted rural life, and stayed in the locality in which they were born.

It was the Industrial Revolution, when machines displaced people power, that led to the massive migration from countryside into towns, and sometimes from far afield. The copper boom of Swansea, the world-wide need for coal and iron from the Welsh valleys, saw vast numbers of workers swarming in from elsewhere, including Ireland. Even after industrial decline set in, many stayed, and it is a tribute to those concerned that they knitted together to form a cohesive community, with a strong sense of identity — but it didn't happen over-night. It took time.

'We are all displaced persons', said a wise old friend. I had been commenting on my Celtic cultural roots and the *hiraeth* which plagues my tribe in its sweet/sad endless way. My compatriots are widely disseminated throughout England — as teachers, actors, politicians and milkmen, but we tend to holiday back in Wales. Home is where the heart is.

Does everyone feel the same? Not all perhaps, but many; and what is their fate in this time of the greatest mass migration of human beings for many generations?

As I write, the European Union has expanded by ten more countries, and we are threatened by the prospect of more economic migrants to add to the current crop of *soi-disant* asylum seekers. While the politicians make encouraging noises, at ground level unease is rife.

The fashionable myth has been of a multi-cultural society. Do I believe in it? I have always had my doubts. True, there is a sense in which Britain is a multi-cultural society: we are a mongrel mix of Celts, Picts, Angles, Saxons, Vikings and Normans, but it has taken us a thousand years to settle down together. Even now the cultural geography of this island is as varied as its geology, which possibly contributes to the high rate of marital difficulties.

The Jewish influx — sixteenth-century Sephardim and nineteenth and twentieth century Ashkenazim — have managed the tension between cultural heritage and assimilation with intelligence and pragmatism, to make a notable contribution to the social fabric. The Indian immigrants expelled from East Africa in the 1960s have found a creative entrepreneurial niche for themselves. The post-war influx of Afro-Caribbeans and Bangladeshi are still struggling to find their identity. It all takes time, a long time, and is often a difficult experience for those individuals caught up in the process.

Is there a multi-cultural society anywhere which really works? The USSR was held together by a repressive regime and fell apart as soon as that repression was lifted. Yugoslavia needed Tito's heavy hand, and since his death there has been Balkan mayhem. Latvia has a large minority of ethnic Russians; some have become full Latvian citizens; the remainder feel disenfranchised and discontented. In Fiji the immigrant Indians outbred the native people, who took legal action to exclude the incomers from political power. The deliberate melting-pot of the USA is no paradise of equality, and ghettos abound there. Less than happy examples multiply.

The psychological reality is that we all feel most comfortable with our own tribe. When with them, we know what to take for granted, what is expected behaviour, where the boundaries are, how to read the faces, when it is safe to relax. In their company, life is less fraught, less anxiety-making, even if it brings its obligations and responsibilities. Unless our background is very dysfunctional or circumstances prevent, we all tend to drift back home in time. We just prefer our own kind.

This is not a moral issue, more a statement about ourselves as social animals. We might manage our national and inter-national lives better if we faced that reality. Too often nowadays I am reminded of a small child at a children's party being pushed to socialise before it is ready to do so. It hides its face in its mother's skirts; if pressurised it screams and has a tantrum.. Developmentally, it hasn't quite got there. Give it time and curiosity will lure it into the group. Force the issue, and the whole socialisation process becomes fraught and emotionally loaded.

The child may remain shy and withdrawn; it may learn to hide its fears by ganging up with the strong, and persecuting others who are weak and fearful. Gangs are an effective way of projecting one's disowned fears into others, and a perversion of true sociability. Rampant nationalism is the same perversion writ large.

We need to be more grounded, less idealistic about these vital issues. We need to acknowledge our innate characteristics as social animals, and not keep trying to force our behaviour into inappropriate patterns which lead to much suffering and are bound to fail.

11 What is out of control?

If we think of politicians' policies and activities as a response to the social zeitgeist, can we perhaps offer another perspective on New Labour's obsession with control and micro-management. Is there a deep unease, a sense that there is some lack of control somewhere in the social fabric that demands attention?

Such absurdities as the minutiae of the Health and Safety legislation which results in a ban on children's playgrounds, and massively inhibits small-scale entrepreneurial activity, reflects a profound insecurity. Why are we such a risk-averse society?

Has the decline of religious faith in an after-life made us more fearful? Or the fading of our replacement faith of total scientific control and comprehension of our material world? This latter illusion fostered a denial of our human limitations, and of our dependence on an uncertain and ever-changing environment. That denial is currently breaking down, and we are faced with the reality that life itself is a risky adventure.

It is a sign of fear and insecurity that we resort to fundamentalism on the one hand, and over-control from the centre on the other. While we immediately think of the Islamic fundamentalists, the American Neo-cons are equally striking. In this country we have our own equivalent among the Evangelical Christian fraternity — the Alpha course — and the Kleinians, no less, in the psychotherapy profession! People in any field who believe that they are the guardians of certain truth are dangerous!

Our children, we are told, are the most unhappy in Europe. A desperate comment on our society! Why? While some no doubt drift into premature sexual activity, drug abuse, binge drinking and so on out of peer conformity, for others these activities are an attempt to manage unmanageable mental pain. Drug rehabilitation programmes which address only the drug problem are bound to fail.

It all begins at home! Many of our young people are inadequately socialised. Social control is reliable only if internalised. It can be learned at the appropriate age, and only from people the child loves and respects. The fundamentals are learned early — pre-school. School subsequently reinforces and develops the learning, but is no substitute.

The mutuality and reciprocity of social life is learned initially within the family. It is begun within the mother-infant duo, then the infant gradually becomes aware of others who have to be accommodated within its world — the father, siblings, and so on. This is not always an easy or welcome adjustment, and the child needs support and reassurance to negotiate it. The active presence of the father (or someone taking on this function) is vital to effect the necessary separation of that first, fundamental close bond; and there is a demonstrable relationship between an absent father for the 2-4 year-old boy, and subsequent aggressive and delinquent behaviour.

I am talking in terms of the traditional family. What happens to children who spend their days in a nursery from a very early age? There is growing evidence that such children are much more self-assertive and aggressive in later years (for good and ill). It is notable that the Israeli Kibbutzim experiment was abandoned after it became clear that the adults it produced were far from being the sane and sound individuals the pioneers had hoped for.

The increasing emergence of rampant individualism within our society is evoking ever-more heavy-handed attempts to control us by legal means. It is a symptom of the same failure of proper socialisation. If social values and attitudes are not internalised — if the development of conscience is inadequate — then those values have to be imposed from without.

It is not at present 'politically correct' to say so, but mothering is a proper job which needs to be well done for all our sakes! The fantasy that it can be done by women who are managing a full-time career, AND competing on equal terms with the menfolk, AND running a home, is a nonsense. A few remarkable women appear to manage it, but close observation usually demonstrates a substitute mothering figure — grandmother, aunt, nanny — all increasingly difficult to come by or prohibitively expensive.

Fathering is just as important, if not as time and energy consuming. Both tasks demand a very long-term commitment. They are different functions and cannot both be carried out by the same person (even if some have to try.) It is a basic and inescapable fact of life that the human young take a very long time to mature!

12 Twentieth century blues

It was in the twentieth century that many present problems began to be more clearly apparent. Virginia Woolf, writing in 1924, said that 'On or about December, 1910 human character changed.' Relations shifted 'between masters and servants, husband and wives, parents and children. And when human relations change, there is at the same time a change in religion, conduct, politics and literature.'

Others too had seen it coming. The Fabian Society,[1] founded in 1884 by a gathering of high-minded intellectuals, recognised the ongoing transformations and set out to develop ideas of socialism.

Of course, these things had been changing for a long time but at that point Woolf had suddenly become acutely and uncomfortably aware of them. Prior to 1914, our society had been subjected to a lengthy and rapid era of change, fuelled by the technology of the Industrial Revolution. Changes in agriculture reduced its need for labour, and resulted in large numbers of people leaving their rural homes to seek work in the towns, and to fill the mills, factories and coal mines. The new towns developed more rapidly than their infra-structure, becoming squalid and insanitary. People were cut adrift from the roots, the price

---

[1] The Fabian Society was founded to encourage gradual, reformist, non-violent social change. It attracted many of the prominent intellectuals of its day, including the Woolfs, and laid the foundations of the subsequent Labour Party.

their labour could command was arguably too low and their living conditions were poor.

The great age of mechanical engineering brought enormous changes, not least in travel, The railway boom opened up the country as never before. By the start of the 20th century we also had the prospect of the motor car and manned flight in view. All this was bound to provoke a reaction; and it did. The Arts and Crafts movement arose out of a rejection of the machine-made, and in an attempt to create for craftsmen a congenial and adequately remunerated life-style.

After 1870, children were subject to compulsory schooling between the ages of five and thirteen. Ordinary people were becoming increasingly literate, they had begun to think for themselves, and their poor living conditions made them restive. They gathered together to form working-men's institutes, and trades unions became increasingly vociferous. They had begun to challenge the status quo. By the end of the 19th century, there was a restlessness around which manifested itself in all directions, including the Arts.

At the same time, the court around the Prince of Wales, subsequently Edward V11, lived a lavish and ostentatious life-style. Within that circle 'sexuality was an upper class sport, and adultery was an aristocratic art-form'. The sharp contrast with the harsh working lives and slum conditions in which so many lived was blatant.

For a long time Westminster had failed to tackle 'the Irish Problem', but by the 1880's desperation had driven a small group of Fenian anarchists to bomb-making. On October 30th, 1883 a bomb exploded on the Metropolitan railway line near Praed Street destroying two carriages and injuring 62 people. In 1884 and 1891 they attacked London Bridge, doing little damage but killing themselves. In different incidents the House of Commons and Scotland Yard were also targeted.

In January 1909, an armed robbery and double murder in Tottenham by two Latvian anarchists fuelled increasing resentment of immigrants.

Then came the battle of Sidney Street in January, 1911 between the Metropolitan Police and a criminal gang of Latvian anarchists, who had perpetrated the Houndsditch murders in the previous month. Winston Churchill, the then Home Secretary, who personally commanded the operation, was criticised by some for a heavy-handed approach; but that affair marked the end of such incidents.

Perhaps it was these events which led Virginia Woolf to her date of December, 1910, as if her defences against the prevailing social unease had finally been breached.

It must have been as unnerving for the people of that time as have been more recent incidents, by different groups of terrorists, in ours.

Is it fanciful to see the period of the Scientific and Industrial Revolution as a time of social hypomania, creative, lively and enterprising, generating optimism and high expectations which the bulk of the population had yet to satisfy? The attempts of the ruling classes to hang on to power misfired when they excluded the Nonconformists from university education. The outcasts took their energy and enterprise where they could, into industry, became successful entrepreneurs, and powered the engine of change.

Historically, at such times, a nervous government will frequently go to war. It is so much easier if one can locate 'the enemy' outside, in another place, than to try and tackle the discontents at home. The Boer War, arguably an act of aggrandisement by the British, acted as a focus for patriotism and provided a unifying activity for the nation.

It also had the advantage of being a long way from home, so the harsh realities of war were largely hidden.

If the 1939-45 war was inevitable, the 1914-18 war was not. There was an arbitrariness about it. Only a short time before we went to war with Germany, there had been serious talk of allying ourselves with the the Germans and warring with France! Was the Great War an attempted solution to the tensions of the time, the point at which the hypomania of the Industrial Revolution burst its restraints?

In the beginning of the 20th century, in Britain many thought that serious intellectual life centred around the Cambridge group known as the Apostles, a self-styled elitist 'band of brothers'. Some of them went on to become leading members of the Bloomsbury Group, and we now remember them primarily for their creativity in the Arts.

They were a mixed group of people. J. D. Bernal embraced psychoanalysis with enormous enthusiasm before abandoning it for his later love, Marxism. Virginia Woolf would have nothing to do with psychoanalysis — since she later committed suicide, perhaps she might have benefitted from what it had to offer. Her younger brother, Adrian Stephen, although feeling himself to be a damaged human being, went on to become a successful and respected analyst.

Among the sounder members of the Group were James Strachey (brother to Lytton) and his wife Alix who together went on to translate Freud's writings into English, and to whom the psychoanalytical world owes a considerable debt.

A recent edition of the *British Journal of Psychotherapy* (November 2010) put me back in touch with the early days of psychoanalysis in this country. In 1924 Alix Strachey went to Berlin for a year to undergo a training analysis with Karl Abraham, the then leading figure of the powerful Berlin Institute of Psychoanalysis. During her time there she wrote daily to James, who remained in England, giving him an account of her experiences; and it is this picture of Berlin life that I found so fascinating.

It comes over as a city throbbing with vitality and creativity at every level. The Arts were lively and experimental, from painting to theatre, from music to literature, from architecture to liberal politics. Intellectual life was buzzing. Alix was taken by her friends to participate in this social whorl — and she loved it. There is an amusing glimpse of Melanie Klein dancing enthusiastically at a succession of balls, dressed as Cleopatra, with a deep cleavage.[1]

Alix was aware of other elements on the fringe; she found the groups of young Nazis in the streets disconcerting and obnoxious, but didn't take them too seriously. As a well-educated English lady who, like Virginia Woolf, admitted to disliking anything proletarian, she probably made little contact with the Berlin world so graphically described by Christopher Isherwood (1935, 1939) amongst others, a world of sexual adventure and perversion, where experimentalism was played out in real life, and nothing was beyond the pale.

The class divisions still operating so strongly in England, had broken down more thoroughly in Germany in this period. There was 'a widespread dream of a better world, a sense of a dawning social utopia, an optimistic hope for a new community of mankind with the general belief in the capacity of the masses to be educated towards the common good.' (de Clerk 2010)

If Berlin, par excellence, exemplified a certain social and cultural trend, it was not unique. London had its Flapper

---

[1] Alix formed a friendship with Klein and was largely responsible for introducing her work into England.

Era, when the well-heeled young partied and night-clubbed to their hearts' content, talking and sometimes behaving as if the old conventional restraints were now irrelevant. The United States had its Jazz Age and the developing movie industry, with its prevalent atmosphere of restless manic activity and dance.

Reading Alix Strachey's account of life in Berlin in the 1920s, I was irresistibly reminded of a more recent phenomenon — that of London in the Swinging Sixties and Seventies. At that time there was the same lively creativity and experimentalism in the Arts, in fashion, in design, in behaviour. There was a similar sense of sexual freedom, the loosening of restraints, the permission to push the boundaries, in relationships as everywhere else. It felt like freedom. It felt like the dawn of a new world.

Whatever happened to all that life? After the Twenties, it all collapsed into the Great Depression of the Thirties, the rise of Nazism in Germany and the dawning inevitability of the 1939-45 war. That in itself produced a major discontinuity in the culture of Europe, with enormous suffering in all directions. The USA did its best to remain detached, was ultimately drawn into the conflict, yet was not as severely affected by it. America provided a haven for many Europeans who made a significant contribution to the intellectual life of their new home, not least in the scientific field.

Post-1945 was a less well defined story. Although the war had officially ended in Europe in May, the war with Japan lingered on, only to be terminated by the release of two atomic bombs, which blew apart a lot of certainties as well as the cities of Hiroshima and Nagasaki and their inhabitants.. Ordinary life in Europe remained basic and bleak for a long time. Rationing in Britain did not end until 1954, and military conscription continued until 1960. The nation was bankrupt and there was little to spare for frivolities, though the Festival of Britain in 1951 was a welcome boost to morale. It took a while for the recovery which allowed the Swinging Sixties to come into being.

With the added bonus of sexual freedom granted by the Pill, anything seemed possible, work was so readily available that for many young people it didn't seem worth the slog of long-term professional training, just as it didn't seem worthwhile to persist with long-term relationships once they became stale. Fashion was great fun, design of all kinds flourished, we built in the International Style ignoring the fact that what worked in the Mediterranean or Californian sun didn't quite stand up to our climate. Youth 'had a ball' with little thought for tomorrow.

It is now 40-50 years on. What happened to all that energy and hope, that liveliness? The descent from the manic high of London's Swinging Sixties has been much more gradual than that of the previous cycle.

There were casualties. As in the Twenties, there had been much experimentation with drugs and alcohol, but what had begun as fun too often became destructive. For some, though not all, the creativity unleashed by drugs fizzled out. Early promise evaporated. Some of the young who had 'had a ball' unexpectedly found themselves middle-aged, with no settled relationships, an awareness of possibilities squandered, and contemplating a dead end in their life's journey.

I want to link the phenomena of the Twenties and the Sixties together as an aspect of the recovery from war. I see them as pathological and ultimately damaging. I suggest they are a form of manic social illness comparable to the manic phase of individual manic-depressive psychosis. Characteristic of both is restless activity, a denial of reality, a squandering of resources, and living as if there were no tomorrow. Attempts at rational restraint are rejected as old-fashioned, small-minded, beneath contempt. There is the delusion of creating a new world, a new way of being in which anything is possible.

It could only end in depression. The 'boom and bust cycle' has once again hit 'bust'.[1]

---

[1] Those whose psychoanalytical perspective is Kleinian will readily see in the above description evidence of the Death Instinct at work.

In his Introduction to Hannah Segal's collection of papers entitled *Psychoanalysis, Literature and War*, John Steiner writes of how 'This hatred of reality and its replacement by omnipotent phantasy is clarified by Hannah Segal in her discussion of the two possible reactions to states of need. One is life seeking and object seeking, leading to an attempt to satisfy those needs in the real world ..... The other has as its aim to annihilate experience of need and the mental pain which goes with it. Here the self or that part of the self that is capable of experiencing pain is attacked, and also the object which gives rise to the awareness of need. Instead of a reliance on reality, the patient turns to omnipotent phantasy as a solution.'

Steiner comments on what he sees as obvious - the 'links with forces of integration and disintegration, anabolism and catabolism, and creation and destruction.'

He goes on to say 'I have always been impressed by the idea that structure and information created in living organisms is the fundamental basis of life and that it is in the process of ageing, illness and death that such structures gradually decay until we return to the dust of inanimate matter. Reproduction, growth and integration can in this way be thought of as representative of the life instinct in a constant conflict with forces leading to disintegration.'

Steiner's latter comments resonate with the current thinking embodied in the Theory of Complexity which suggests that living organisms have an in-built tendency to spontaneously develop ever more complex forms of organisation. One could interpret this as the Life Instinct at work.

13 The perverse society

This post-Sixties period has been dominated by a phenomenon which was blatant in Twenties Berlin, but which also existed elsewhere during that time — the phenomenon of perversion. (I am using this term in a technical sense, although it does embrace the common interpretation of sexual perversion.) I suggest that we have been living in a perverse society, using the term *perverse* as defined by the French psychoanalyst Janine Chasseguet-Smirgel.

According to Leigh (1979), 'Chasseguet-Smirgel ... thinks that perversion constitutes a universal temptation of the human mind. This temptation is to attack reality by attempting to dissolve or deny all limits and boundaries and all differences. It wishes to rise above the ordinary laws of time and space. This process of confusion, of denial of specific attributes and the attempt at homogenisation becomes idealised and this forms what she terms the anal universe. The attack on separation, differentiation and naming, on specificity, leads to the construction of a substitute reality where anything can become anything else as a defence against the pain of having to accept one's relative place in reality. This is something that everyone must struggle with, that is, the giving up of narcissistic omnipotence and the maintenance of illusion, as well as the wish to own and control and use at will one's objects.

The perverse act therefore seeks to create a new reality. It is also an attack on the mind and the functioning of the mind and on thinking... To use Bion's term, this is an area of -K, where there is a breakdown of perception, knowledge and awareness.'

- Under the guise of good liberal words and notions, I believe that, for decades, we have been engaging in a flight from reality. Under the guise of equality we have pretended that all people are much the same.
- Under the guise of equal opportunity we have refused to recognise that different people have different levels of intelligence, different innate abilities, different gifts, and that these need to be met by appropriately tailored education.[1]
- Under the guise of feminism, we have pretended that men and women are only made different by social conditioning..
- Under the guise of ageism, we try to deny that people do change over the years, for good and ill; that experience has value, and that declining physical resources need to be accommodated. (How many of us are truly capable of full-time work until 70?).
- We treat pop music as of equal value to the classical European canon, and in that connection raise to iconic status narcissistic personalities with minimal talent.

---

[1] We have pretended that by changing CATS (Colleges of Advanced Technology) into universities they have achieved the same educational standards as the older universities. We have changed almost all our schools from streamed to comprehensive with a 'one model fits all' approach, and now wonder why we are woefully short of skilled craftsmen, engineers and scientists. We have discouraged competition between pupils, individually and collectively, ignoring past experience that boys, if not girls, love competing, and that this appeal to the male pecking order may be the only way to motivate them.

- Finally under the rubric of post-modernism, we put forward rational arguments that there is no ultimate meaning in anything, thus wiping out history, theology and all that experience and tradition has given us.

Chasseguet-Smirgel writes of how the pervert, in his inner world, reduces everything to an anal mess and muddle, like a child mixing together all the colours in its paint-box. The end-result is not the glory of the rainbow but a dirty formless mud. It is no coincidence that the Turner Prize of 1998 was given for an exhibition of elephant dung, or that other recent prizes have been for a pile of bricks, an unmade bed and the embalmed half-carcass of a calf. As always, our artists have captured the spirit of the age.

I noted, too, that in August 2001, Channel 5 television put out a programme in its *Real Sex* series on 'Whip-cracking sessions and a pageant for older strippers and dancers'. What would Lord Reith[1] have said!

The IPA report[2] goes on to argue that while the social acceptability of certain kinds of behaviour does vary at different times and between cultures, 'acceptable' does not necessarily imply 'harmless', either to the society or to the individual. To assert otherwise is to indulge in just that homogenisation to which Chasseguet-Smirgel alludes.

I believe that this social perversion has spread to all areas of our cultural life. It is manifest not only in the Arts, but also in Science — the willingness to contemplate creating human/animal embryos; in political life — the expenses scandals at Westminster; in our financial institutions with

---

1 Sir John, later Lord, Reith was the first Director-General of the BBC. He did an excellent job for his young organisation in many ways. However, he was a Highland Scot from a Calvinist background and of a somewhat dour temperament, and his firm hand on programme content reflected his own perspective.

[2]*Panel Report on Perversion* from the Barcelona Congress 1979 of the International Psycho-Analytic Association. *See* Leigh R.

their lack of judgment around debt and sub-prime mortgages; in the very notion that the financial world, with its talk of 'financial products', has the same concrete reality as manufacturing, engineering and agriculture.

Perverse organisations

Susan Long (2008) outlines what she sees as the characteristics of a perverse society. Perversion, she argues, is not just a deviation from normality but is about the seeking of individual pleasure at the expense of others.

The perverse state of mind acknowledges reality but at the same time also denies it. Since reality threatens both self-interest or previous certainty, so a perverse organisation promises the creativity of holding opposites in mind, but their connection is illusory. It offers also the destructivity of denying truth or reality and instituting parasitic relations with others. The simultaneous acknowledgment of reality and its denial is the lure of the perverse.

The collective human need for certainty in the face of threats to survival is primitive and pervasive. This need for certainty gathers people together as accomplices and shapes what humans can collectively tolerate.

Certainty requires confirmation by another, Long argues,. In order to confirm the perverse stance, accomplices are needed.[1] The perverse state of mind therefore engages others as accomplices—conscious or unconscious—in the perversion. The accomplice is treated as an (autistic) extension of self and, she believes, tends to develop a parasitic relationship.

The perverse mind-set flourishes where instrumental relations have dominance in society, because

---

[1] Rorty (2008) argues persuasively that truth is always a social construct. We apply the label to the consensus view of those we judge informed about the issue in question, and have in reality no access to any absolute truth, whatever that might be. See also Gosling (2001).

instrumentality masks the more extreme issue of abuse. One could interpret the exploitation of the working classes, including domestic servants, during the development of the Industrial Revolution as abusive. In both cases the relationships were essentially instrumental.

Organised corporate corruption is a conscious manifestation, the iceberg tip of of an unconscious perverse societal structure and dynamic. Corruption builds on an underlying social fabric of perversity. Perversion begets perversion. Abusive cycles are hard to break.

While Long is writing about our contemporary culture, her comments apply with equal aptitude to Edwardian society in England. Perhaps it needed the 1914-18 war to break the abusive cycle.

The psychoanalyst Thomas Ogden (1989), writing about psychic activity in the individual has made a unique theoretical contribution which the writer has found helpful in the consulting room. He suggests that different modes of psychological functioning belong to different stages of emotional maturation. These he names as the autistic-contiguous, the paranoid-schizoid and the depressive, and he believes that these different modes of functioning co-exist and remain operative within the adult psyche.

He sees the autistic-contiguous mode as developmentally that of the infant whose experience of the world is through its senses. Subsequently this mode continues to provide the sensory floor of experience.

The paranoid-schizoid mode is developmentally that of the small child with its immediacy, its characteristic splitting of experience into good/bad, me/not me, and its concrete way of thinking and expressing itself.

The depressive mode is a later development and with it comes the capacity to think symbolically, to contain feelings, to wait, to be aware of one's separateness. Only in this mode is one fully aware of the Other as a separate and autonomous entity. It is also the mode in which we have some historical perspective, because while in it we have a sense of the passage of time.

In Ogden's words 'Experience is always generated between the poles represented by the ideal of the pure form of each of these modes.'

He suggests that psychopathology can be understood as forms of collapse in the equilibrium between these three modes. Collapse towards the autistic-contiguous pole and we become trapped in 'the machine-like tyranny of attempted sensory-based escape from the terror of formless dread, by means of reliance on rigid autistic defences.' Collapse towards the paranoid-schizoid pole and we become trapped in 'a non-subjective world of thoughts and feelings experienced in terms of frightening and protective things that simply happen, and that cannot be thought about or interpreted.' Collapse in the direction of the depressive pole involves 'a form of isolation from one's bodily sensations, and from the immediacy of one's lived experience, leaving one devoid of spontaneity and aliveness.'

While he is here writing about the individual, I suspect similar processes are at work in the wider social scene. One could perhaps understand the 19th century Evangelical Movement in Britain as society's collapse in the direction of the depressive pole. We have only just recovered from 'the isolation from one's bodily sensations, and from the immediacy of one's lived experience' that accompanied that phase. As a society, having in the last fifty years recovered our spontaneity and aliveness, none of us would wish to go back to the inhibitions and repressions of the late nineteenth and early twentieth centuries.

My sense is that recently we have been tending towards collapse in the direction of the autistic-contiguous pole, desperately seeking sensory-based satisfactions, while at the same time many of us choose to live alone because that is the best we can manage. We get together with others to 'have fun' or sex, but spend much of the rest of our lives relating to a computer, a television set or driving vast distances in cars. Perhaps the current ubiquity of the mobile phone is a plus factor after all. It is human contact of a kind, if only at a distance.

What Freud termed polymorphous perverse sexuality—that very early, unfocused expression of sensory excitement in the normal infant—would seem to belong to that phase of development when we move from the autistic-contiguous mode into the paranoid-schizoid one.

Is one element in the generation of perverse behaviour in the adult a regression back to that primitive level of experience? Is there a danger, if perverse behaviour becomes less acceptable, that we see the emergence of social paranoid-schizoid phenomena such as characterised revolutionary France and Nazi Germany?

14 The perverse society in terminal throes

The narcissism and individualism existing in our society's late twentieth-century establishment has, through its values of self, greed, consumerism, acquisition and exploitation promoted the emergence of perversion through the process of turning a blind eye.

I see all this as the long-term fall-out of war. The terrible loss of life in the 1914-18 war, the appalling conditions which the soldiers had to endure; these were dreadful enough. The void left by dead lovers, dead husbands, dead fathers; the deprived women bereft of their hopes and expectations, forced into a working life they had not chosen, the collapse of great estates because there were no heirs left to manage them; this was what we were left with

I believe it was all too awful to contemplate, to face, let alone begin to assimilate, for a very long time. Manic denial allowed people to go on living as best they could. Without it the young would have been too weighed down by the depression and suffering of those around them. It protected them from what might otherwise have destroyed them.

It was a similar story post-World War II. In addition to the usual horrors of war, we have had to come to terms with nuclear weapons and all the implications that they bring with them. It has been mind-blowing. I feel it is only recently that we, as a society, have been able to get some perspective on what has happened, to begin to emerge from what is, in social terms, a pathological regressed state of being.

I do believe that we are beginning to emerge from this phase. The excesses of the early feminists have been rejected and modified by their daughters and granddaughters as the realities of childcare have impinged..

There is a growing realisation that our education system is not delivering what was hoped for, and there are moves afoot to change it while insisting, for political reasons, that we are not turning back the clock. There is growing evidence that an approach which suits girls very well does not give boys what they need; that in a women-dominated profession, male teachers provide something much needed and valued.

Similarly, in the field of dietetics which has been a niche traditionally filled by women, we are beginning to realise that men and women thrive on distinctly different food— one which seems to reflect the Stone Age human specialisation into male hunters and female gatherers! Women love salad, but 'real men don't eat quiche!'

In a recent radio programme contributors protested against the commercial pressures towards premature sexualisation of young girls[1] and suggested legislation to curb such exploitation. In doing so they were echoing an unease put forward some time previously by Rowan Williams (2000) in his book *Lost Icons* He saw the problem as the erosion of boundaries around the latency period of childhood, arguing that that peculiarly human stage of development was essential for the proper emotional maturation of the young person. Again, we have

---

1 Including padded bras for 7-year olds!

witnessed a cultural denial of difference between the child and the adult.

One could link, perhaps unkindly, the perverse spirit of our age with the cultural dominance of the USA since that nation was founded upon denial—from its basic tenet that all men are equal, to the expectation and encouragement that its immigrants forget about their roots, their countries of origin. While there was a certain wisdom in these attitudes in that they enabled a gathering of very disparate people to cohere into a nation, nonetheless it denied certain basic human realities. Am I reading too much into the immediate present to link the beginnings of change in our own society to what I feel is a decline of American dominance in the world, and the current rise of Creation Theology[1] in their thinking?

It was bound to happen sooner or later. Fritz Perls' Gestalt mantra of the sixties—'you can become anything you want to'—had become 'you can have anything you want', and we were living in such a way as to actualise it. The average cost of a wedding in 2011 was £20,000, hen and stag nights expanded to become weekends; some even became weekends abroad in Thailand or the Caribbean.

Debt had become an acceptable way of life, sanctioned by the state, which was content to see young graduates beginning their adult lives with debts of many thousands. Second mortgages were raised on homes to fund holidays or a new car, fashionistas bankrupted themselves to acquire designer clothes, while the gas-guzzling 4x4 USV ceased to be a working tool and became a suburban status symbol, the 'Chelsea Tractor'.

---

1 Creation Theology is currently widespread in the US, especially amongst Christians holding fundamentalist beliefs. It involves a denial of Darwinian evolution, and explanation of the pre-history of our world in terms of special creation by a Deity.

A young academic earning an adequate but far-from-lavish salary felt like the poor relation amongst her school cohort, many of whom had moved, post-university, into the financial sector. To her, paying £70 per week for dog-walking fees seemed somewhat excessive the day after she had been interviewing a woman in the Welsh Valleys who was struggling to do three jobs in order to maintain her family. There were those who had missed out during the spending spree. They had become an invisible embarrassment.

How was it all going to end? What we hadn't foreseen was that the widespread lack of balance and discrimination had spread even into our banking fraternity. Who would have thought that the perverse denial of difference had impaired those shrewd people into not seeing the difference between a realistic mortgage and a hopeless one. The lust for limitless wealth, and a manic lack of responsibility, had impaired their judgment to a mind-boggling degree.

The bottom fell out of the market with a bang. Judgment Day was at hand. Never again, we hope, will the bankers have the freedom of manoeuvre they have enjoyed as the state steps in to pick up the pieces and regulate their behaviour. The demand is that limits and conditions be firmly imposed.

This absence of limits and boundaries is another symptom of a perverse society. It can be exhilarating. It can stimulate creativity—at least for a time. After a while, the urge not to do anything which involves struggle does tend to undermine the perception of reality. It is a bit like the effect of LSD—a surge of creativity followed by burn-out. Any sense of a hierarchy of values tends to fade. All relationships are of equal value so why bother to get married or work your way through a bad patch. Why learn the techniques of drawing and painting when people will pay vast sums of money for an unmade bed or a dissected calf. An imaginative montage may have meaning, but a random collection of debris does not, except perhaps to the schizophrenic who has destroyed any meaning within his inner world.

Where now? Perverse behaviour was rampant in the late days of Rome, before the barbarians flooded into the Western half of the Empire, and the centre of power moved to Byzantium. It was also widespread amongst the upper classes in the late Georgian and Regency period, and it would seem to be in reaction to such behaviour that the 19th century Evangelical Movement took such rapid hold. It was a feature of the Weimar Republic if Christopher Isherwood's lightly fictionalised account is to be believed. Would the Nazi ideology have had the appeal it did except in reaction to that which preceded it?

A certain authoritarianism has been creeping into our political and judicial system, and we have allowed it because the alternative—a paranoid-schizoid world of anarchy and destruction which terrorism threatens to create—feels like a worse alternative. Perhaps some corrective is necessary, but the trend has its dangers. Historically, at such times an attempt at stabilisation has evoked the emergence of a 'knight on a white horse', a strong man, who all too often develops into a dictator. Just within the twentieth century Hitler, Mussolini, Franco and Salazar could all be seen in this light.

Where are we now? Recently we have been living through the great age of electronic engineering. It has brought us a revolution in communication — mobile phones, ever cheaper telecommunications, Skype, and, above all, the inter-net. For good and ill, privacy is invaded, secrecy almost impossible. One suspects that Wikileaks will have changed diplomacy for ever. What will happen to libraries and publishing as e-books replace the printed page? When up-to-the-minute news is available on numerous television channels and the inter-net, can newspapers survive?

It is all happening so quickly, and it is difficult for the less flexible amongst us not to get left behind. It is bringing change at an unprecedented rate.

15 Where next?

I sense that our European society is once again in transition. Reality is seeping in, and uncomfortable it is in many ways, but it demands to be addressed. Retail therapy is losing some of its appeal as the cost of housing weighs heavy, and the pension crisis intrudes. The threat of climate change can no longer be ignored. Diversions become ever more difficult to find — or to justify.

I have no slick answers, only questions. I am just noting a certain pricking in my thumbs, an unease as I scent possible dangers ahead. "Truth is the daughter of time", but the timespan of a human life is short, and long-term trends are obscure. I should like to see the underlying phenomena better understood so that we can hopefully avoid further self-destruction in the future.

So then, what next? I have long suspected that the inevitable next phase of society is a depressive one. I just hope, looking at the past, that it is not a paranoid-schizoid one. The politicians have learned something — hopefully — from the Great Depression of the Thirties, and are unlikely to let the full horrors develop again.

There have been some signs of impending change for a while. Clothes designers have been looking back to the Thirties for inspiration and Art Deco has become very fashionable. That era communicates a depressive mood to this observer, who did not much like it first time round.

Undoubtedly we shall see the imposition of many more limits and boundaries. Money will be tight, retail therapy less of an option; thought will, of necessity, precede impulse.

Will our sexual morality change? We have made gains in that dimension few would willingly lose, but again maybe some sense of appropriate boundaries would be welcome. A recent furore at the BBC, centred on a tasteless broadcast by Jonathan Ross and Russell Brand, suggests that public opinion has reached the limits of its tolerance. I suspect drug abuse may lose its dangerous glamour; so, hopefully, will binge drinking among the young.

The re-emergence of patriarchy is a prime danger at such times. Reassertion of the importance of fathering is to be welcomed, as is the re-gaining of a proper sense of male pride, but how will that impact on young women as they juggle their aspirations of career and family life.

The old skills of make-do-and-mend are reappearing. The art of cooking from basic ingredients is being promoted in television programmes. Multi-tasking is now a recognised skill in which women excel, and they will need it all if they are not to lose painfully gained ground. Many of today's young women have the self-esteem needed to be assertive, but what about their daughters in years to come? The flighty young women of the Regency period became the respectable matrons of the Victorian era. They must have been, to some degree, tongue-in-cheek given their own youthful behaviour, but their daughters took it seriously.

We may experience a contemporary version of the Evangelical revival but some aspects of it we emphatically do not need. Our society has only just recovered from the last bout.

Well, time will tell and I shan't be around to see much of it. It will be tough for some individuals but we may develop a saner society out of the mess. We can only hope.

Three years on from the world banking crisis of 2008, and we are struggling on a depressive plateau. It has become apparent that during the years of plenty, while the rich got (a lot) richer, the poor and the moderately affluent did not. Their situation remained largely unchanged. The myth that wealth trickles downwards was revealed as indeed a myth.

While some attempts have been made to curb the power of the banks, they still award themselves enormous bonuses while being very unwilling to lend money to those requesting it for mortgages or small businesses. So far, in spite of general opinion, they have resisted attempts to separate their retail business from their financial gambling activities. Few politicians seem willing to confront these issues.

In the meantime, Europe is struggling with the financial debts incurred by Greece, and Italy and Spain, which threaten the stability of the Euro; while the USA similarly struggles with its chronic overspend, and the impasse between Democrats and Republicans as to what to do about it. The tendency for the powers-that-be to recommend economic stringency is seductive, but tending to diminish economic activity and hence leading to further depression.

However, in Britain another major factor has changes our calculations, with the exposure of the newspaper hacking scandal. Has the Murdoch empire has been pulling political strings behind the scenes for a long time? Has it been unscrupulous in the methods it has used, literally guilty of corruption of the police force, and virtually of our body politic? For the moment these are open questions, but the answers, whatever they are, will mark the end of an epoch. That period has demonstrated those typical diagnostic signs of a perverse state of affairs in the blurring of boundaries, and the denial of ethical standards. It has been a case of 'anything goes in the service of business' whatever the cost to the individual or the social body. Whatever the final outcome, one feels that the power of an unconstrained Press can never be the same again.

The truth is that we have been living beyond our income for a long time, and the day of reckoning is dawning. We have squandered so much of our resources — both material and human — while neglecting the basic underpinning of our social life. Even the USA, with its vast resources, is currently the member of the international community most deeply in debt, and is perhaps realising that it can no longer afford unwise military incursions of dubious value and intent.

We have a major task ahead of us. How do we hold on to the achievements that technology has brought us while living within our means? Unless we can do this we shall destroy our world, or — a likelier scenario — it will destroy us.

Freud stated that the sign of health was the ability to love and to work. Work provides livelihood, structures time, hopefully allows an outlet for (and stimulates) creativity, gives status and defines one's role in society. Love is the basis of good relationships, within the family context, in face-to-face groups and in the wider society.

The innate human disposition seems to be to live in relatively small face-to-face groups. Once we operate in larger groups, social organisation becomes political, open to abuse, less meaningful to the individual. 'Socialisation' — the internalisation of standards acceptable to the wider society — becomes more imperative, but more difficult to achieve. The alternative is alienation, exploitation of social resources, criminality. So how do we facilitate the individual's ability to identify with the wider society?

Our society needs to retreat from its present position.

- It needs to live more locally. Globalisation is a disastrous fantasy, which denies the reality of cultural difference. It allows the unscrupulous to exploit the rest — as has happened in the recent banking crisis.
- It needs to value its communities where they still exist, and have not been destroyed by thoughtless 'development and improvement'. The profit motive, though indispensable, needs to be firmly constrained by an enlightened framework of law, and should not be paramount in public services. Unbridled accountancy is destructive.
- It needs to hang on to small-scale institutions: its village schools, its Post Offices, its small shopping areas, its local hospitals.
- We need to encourage extended families to stay close to each other, after a period of massive social and geographical dispersal, which has left vulnerable the young parents, the elderly, the less than fit.
- We need to find a mechanism for socialising the young adult. University education does it for those who achieve it, though it is arguable whether it is appropriate for all those to whom it is offered.[1]

The internet is the determining technological advance of our time, and it has proved a mixed blessing. It allows us easy and rapid communication — wonderful! However, it is not a total substitute for face-to-face encounter. Even Skype has a limited capacity for communicating bodily

---

[1] I favour some kind of civilian equivalent of the former National Service. If handled appropriately, it could provide the young person with a mixed social group which could continue in existence subsequently. It might also teach useful skills — there is plenty of community work currently not done. By taking individuals out of the stultifying, hopeless 'sink estates' in which they are trapped, it could show them other possibilities. Perhaps it might even break up the gang culture of our inner cities.

tension and gesture, and it is devoid of smell and touch. We rely more than we know on these subliminal perceptions in forming judgments of our fellows.

The internet denies the reality of space and distance. It encourages us to live in a fantasy world — some of it malevolent. Where there is no reality check, human relationships easily degenerate into instrumentality, and then perversion flourishes. The retreat from personal encounter — in the public libraries, in the banks and many other aspects of social life — in favour of reacting with machines, isolates us ever more. It is to be regarded with profound suspicion; guilty until proved innocent.

We need to look carefully at those small scale societies which seem to work — among them Switzerland, Norway and Wales. Switzerland seems to have solved the difficulties of weaving disparate ethnic and language groups into a cohesive whole. It has done this by keeping central bureaucracy to a minimum, and allowing the separate cantons maximum freedom. If the end-result is a somewhat over-disciplined mind-set, the Swiss clearly think it preferable to the petty warfare of the distant past.

Norway apparently has all the virtues of Scandinavia without the depression and alcoholism of the Swedes. They seem to enjoy life in spite of their climate and limited natural resources! It is a notably egalitarian society, with a high standard of living maintained by high taxation, which appears to be well tolerated. There is a lot of evidence now that the more equal the society, the happier. The Norwegians seem to be a good example.

Wales has the virtues of a small, essentially face-to-face society. When a group of people meet together almost their first activity is to tease out their family and social connections! The clan organisation of old is still operative.

Everybody knows everybody in terms of being able to place them.

The positive aspect of this is that the Welsh have been able to resist the onslaught of English culture in spite of six centuries of occupation. The Welsh are still a different people with different attitudes and values. They have a feeling for the collective, for the cooperating group, which is very different from the competitive individualism of the English culture. Their variety of socialism is communitarian.

They have a feeling for the family and for holding it together. They actually like children, and encourage their development to an impressive degree. The downside can be a certain lack of initiative derived from the need for family solidarity — perhaps inevitable if survival has depended on staying together. When they do fall out, they do so dramatically! Paranoia becomes rampant. Clan cohesion can lead to something that from outside looks like corruption — as in other parts of the world.

All three of these countries have a relatively low population density. Large conurbations are few. The landscape discourages urban sprawl. Mountains provide natural barriers. This may be relevant to their achievements.

Historically they have each had a powerful Protestant ethos which has encouraged social responsibility while battling difficult external conditions. None has grandiose global ambitions. They are not hampered by past experience of Empire. They can focus on what is needed *now,* and what is realistically possible. They are good places to live. What can they teach us?

There are straws in the wind. There is a rapidly mounting demand for locally produced food in England. Farm shops are flourishing. The ever-rising cost of energy is also having its impact. Hopefully it will result in less travelling, substituted by more internet contact within businesses.

In the second decade of the twenty-first century the energy challenge is beginning to occupy people's minds. The government is giving grants for roof insulation and solar panels. Small water-mills are being brought back into use. Already there are currently three on the River Frome in Somerset. For both of these renewable energy sources, excess power generated is sold back to the grid on favourable terms

The virtues of localism are beginning to be perceived. Tory leader David Cameron is looking to return political power from the centre to local communities — schools and so on. He is also talking of the possibility of short, voluntary periods of community service for teenagers. Could this be a precursor for a more extended and developed scheme?

The Liberal Democrats are promising to stop the building of more prisons, and a move to using other ways of dealing with minor offences. This could be an opportunity to further develop the concept of Restorative Justice. It is ridiculous that we have the highest prison population in Europe *pro rata* to our population — yet it doesn't work! Of first offenders, 90% re-offend.

While church membership is still in decline, there is an upsurge in spirituality of many kinds, as people search for a *Weltanschauung* which can give them a framework for creative living.

These are straws in the wind, no more, but they are encouraging. However there is a larger question. Do they portend a move away from the perverse form of social organisation? We need to keep our feet on the ground, turn our backs on fantasy, be aware of our ultimate total dependence on our environment, and be sensitive to the humanity of each and every member of the human race.

A tall order! Perhaps what we all need, as we have always needed, is a perspective, a faith if you like, that will carry us through the dark times and the good. Is such a thing possible in our modern secular world? Shall we survive without it?

16 New world, new language

If the Scientific revolution has brought great benefits, it has also verbally impoverished us all. Inherent in its process is the impoverishment of language and concepts. This is done quite deliberately to facilitate the development of ideas. For example, *energy* has come to have a very specialised meaning in physics, as has *information, momentum, potential* and *induction*. By stripping away a broader meaning, events can be subjected to mathematical analysis, and understanding is deepened. Out of the deeper understanding comes new possibilities and opportunities.

This stripping away is acceptable as long as it is understood for what it is — a technique, a tool which we use in order to create models of how the world functions. What has all too often happened is that we have been seduced by our own cleverness into believing that it is more than a technique, that it expresses a true reality. In falling into that error, so much has been lost on the way. The associations are lost, the nuances and subtlety; the poetry is lost, the wholeness of experience. Life, as we live it, becomes increasingly fragmented and disintegrated. The patterning, the order, which we take for granted, and which gives us much of our inner security, no longer holds.

The limitations of words

Words are a mixed blessing, as those in the public eye know to their cost.

- Words can communicate, or they can distort.
- Words can illuminate, or they can confuse.
- Words can confirm us, or undermine us.
- Words can heal, or they can wound.
- Words can set us free, or they can imprison us.

Some of the most formative events of our lives, which go to make us who we are, happen before ever we have words. There is a lot of firm evidence now that our individual life within the womb is a variable experience and contributes to our subsequent mind-set. Undoubtedly, the world we enter at birth matters, whether it is one where we are welcomed, cared for, kept warm and safe and handled with sensitivity; or whether we arrive to a painful emptiness, are treated as a thing, left alone for long periods feeling vulnerable and helpless — all this communicated through basic physical sensation. To recall those early weeks and months takes a long period of deep therapy and an experienced therapist, but when one gets there there is no doubt about it. It is a transformative experience.

But what of those other transformative experiences which are beyond words — those experiences of the numinous when we touch the fringe of that infinite mystery some of us call God? They can be as powerful, as creative, as transformative as anything that ever happens to us, yet they are utterly beyond words.

Even so we do try to find words for them — to capture the experience, to hold on to it and hopefully to share something of it with others. However, those who hear the words often fail to get beyond them, giving them a reality, a concreteness never intended, building systems of thought around them which bear little relationship to the original experience, and becoming all too attached to their own wordy creations, using them to defend against uncertainty and doubt.

Then they meet up with other people who have used different words, and find themselves quarrelling with them, fighting for forms of words which enshrine their own certainties as if for life itself. From a detached viewpoint you may wonder how grown people can be so foolish. In truth, the ability to live with uncertainty and doubt is a task of maturity. However, existential anxiety[1] is an inevitable accompaniment of these challenges, and our capacity to bear it is both variable and limited.

The authoritarian regimes which plagued the twentieth century arose in societies which were in a state of chaos and disintegration. We cannot blame scientific thinking for creating those conditions — though they were often used to justify them — but it was in reaction to those conditions that the regimes took hold.

Perhaps we should understand Marxism as a consequence of existential anxiety, a failed attempt at defence by applying scientific-type thinking to social questions. The resultant reduction of human beings to economic units, seeing them as undifferentiated entities within a rigid social class, says it all — and has been massively rejected nowadays by most people. However, for a time it clearly gave its adherents a unitary vision, a sense of empowerment, and protection from uncertainty and doubt.

Today, if we suffer as a society — as we assuredly do — from bureaucratic centralism and an excess of controls, is this perhaps a desperate attempt to hold everything together, a force of words to set against that threat of total

---

[1] Defined by Paul Tillich as arising when a being becomes aware of the state of non-being.

disintegration which leads to social chaos and personal madness?

Yet some means of defence against existential anxiety there must be. I am be grateful for my own Quaker tradition which has always valued silence above words; grateful also for those others with a talent for words who enable us to share their experience, in however imperfect and partial form, both now and down the ages. As individuals and as a society we need a *Weltanschauung*, a wider perspective which serves to contain our anxieties and fears. We need a religion by whatever name.

17 Contemporary religion

Are science and religion necessarily enemies? Anxiety about the apparent confrontation is not new. Ralph Waldo Emerson (1803-82) gave shape to the anxiety in a famous quotation: 'What terrible questions we are learning to ask! .... We are coming on the secret of a magic which sweeps out of men's minds all vestiges of theism and beliefs which they and their fathers held and were framed upon.'

Yet the often-voiced opposition between science and religion will no longer do. In our contemporary society, any religion with serious claims must somehow encompass and accommodate our current scientific understanding of the world.

I have never understood the often discussed conflict between religion and science. Science teaches us much about the way the world functions, the 'how' — but not the 'why'. This is not in any sense to decry science. It has brought us many benefits, enlarged our understanding, opened up immense possibilities. If at times those possibilities have been used for destructive ends, that is not inherent in science; rather a statement about human choice.

Formerly, each of the two camps, science and religion, has loudly insisted that they, and they only, hold the Ultimate Truth. Certainly, many scientists enter the discipline driven by a search for truth, and spend a creative lifetime in that search. However, there is a developing understanding that all science can ever provide is a series of models of how the universe functions. The value of these models is that they have predictive capacity. However, when, as happens from time to time, the models do not quite fit the always growing understanding, then a new model is developed. This is the history of science. (Kuhn 1970)[1]

It is interesting that scientific models sometimes have great predictive value even when they are known to be wrong. The work of James Clark Maxwell on electromagnetics in the nineteenth century is a striking example. His concept of the luminiferous ether, on which his theory depended, was out-dated within a few years of being propounded. However, his equations have continued to be taught and used because they enable the solution of some very difficult problems. Errors arising from the initial mistaken physical assumption are mostly too small to be significant.

There are those now who argue that a quantum model supersedes these equations, but that is difficult to accept by those who have struggled for years to understand and use them! The level of passionate resistance to a new quantum model demolishes any lingering belief that all scientists are creatures of reason!

In turn, religions offer models on another scale — models that attempts to describe the ultimate meaning of the universe. The convinced believers of each religion tend to be certain that they are in possession of the Truth about the Universe, and that everyone else is wrong. Derived

---

[1] It was Dame Kathleen Lonsdale (1903 - 71) the distinguished crystallographer who wrote, 'I have learned as a scientist how much I don't understand. I have leaned too that when a scientist encounters two apparently irreconcilable ideas, these are the stepping stones to new knowledge.' 1962

from these primary models, we have further models of how to live the God-fearing life — the Judaic Ten Commandments, the Christian doctrine of Love, the Islamic code of observance, ritual and so on. Although differing in emphasis, all these codes of conduct have much to recommend them if faithfully followed. Sadly the reality all too often fails to live up to the aspirations.

Historically, human beings have fought wars on these issues, have persecuted non-believers, tortured them, burned them at the stake, blown them to smithereens, excluded them from aspects of public life, and generally behaved in ways that were a disgrace to any religion. If some people reject all manifestations of religion, one can understand why.

The Ultimate Truth is surely that we cannot know it. It is beyond our comprehension. We simply do not have the brain capacity to envisage it. We may get glimpses of it from time to time. We may intuit it. Certain people seem to be more gifted in this way than others, and the founders of the great religions are in this category. They have each achieved an insight which has blown their minds. They have then tried to communicate it to others, using words as best they could. Inevitably their attempts at communication are inadequate — how can words contain the ineffable! The tragedy is that those who hear them them codify what they think they have heard, and the system subsequently becomes set in concrete. The spirit is lost.[1]

All revelations are partial; all revelations are inadequate; all are limited by the human personality of the individual concerned, and by the language used, which is of its time and place. All have something to contribute; all are worth consideration, but... none have the Ultimate Truth. The

---

[1] 'The Spirit quickeneth but the letter killeth.' George Fox

High God is unknowable; the Hindus acknowledged that many centuries ago. If we could accept that basic insight, we could respect each other's traditions, learn something from them, begin to fashion a new formulation appropriate to our own time and culture.

Because of the pressures of population growth, the development of world-wide communication networks, and the recent massive migration of peoples, the tensions are growing. We can no longer isolate ourselves within our own cultural box; we have to find accommodation to other faiths and traditions; we have to assimilate other perspectives, including the scientific one. Perhaps the scientific one can be our salvation since it is a world-wide culture which crosses the old religious boundaries. [1]Islamic science today is indistinguishable from the Western variety. The laboratories in Teheran are as sophisticated as those in Cambridge or MIT[2].

Why then are we seeing, not a coming together of religious perspectives, a softening of the boundaries, but a rise in fundamentalism, with its accompanying hardening of the differences?

Sure and certain hope?

An important function of religion, all religions, is to help those who practise it manage their anxiety. Religious belief addresses that fundamental, primitive, existential anxiety. It addresses it by offering an explanation of why the world is as it is. Frequently it offers hope for a future life after death. It provides a community of fellow believers who offer support and encouragement. It offers sacred ritual, codes of belief, codes of behaviour, all of which act to contain the anxiety inherent in being human, being alive

---

1 Yes, there are those who reject it in favour of a creationist myth, but they are relatively few.

2 Massachusetts Institute of Technology

in an unsafe world. It offers comfort, stability, and a sense of identity to those who belong.

However, our contemporary world, while mitigating many age-old fears around hunger and health, in other ways adds to our anxiety. Our sense of identity is ever more difficult to establish and is continually under threat. Mobility — social, economic, geographical — leaves us ever more rootless. Extended families become increasingly dispersed. Any sense of community is difficult to find.

Also, for several decades our society has been living through a perverse phase — perverse in the technical sense. It has been pervaded by a denial of difference, a blurring of boundaries, a questioning of standards, an erosion of hierarchy, an abandonment of rules. It began as something creative; it has left too many people floundering and at sea.

For those who migrate into a foreign culture, the task is more difficult still. A fundamentalist faith can rescue us from inner disintegration, as we are noting amongst second-generation Muslims in Great Britain. In the Judaic faith, while the Liberal and Reformed members are tending to get assimilated, there has been a notable increase in the Haredim — the strict Orthodox — in recent years. The rise of the religious right in the United States has made it a political force to contend with.

A religion with clear rules, firm boundaries, can more readily restore our sense of significance, of identity. It can give us permission to shut out the constant undermining of our well-being by the overwhelming flood of information and bad news emanating non-stop from the media. It can help us protect the inner core of our being. Above all, it can give meaning to our lives.

To succeed, any new religious formulation must do all these things — a tall order. Outside a fundamentalist faith, and since Communism collapsed, the only other conceptual framework which comes anywhere close is the scientific one. Yet something odd is happening nowadays at the cutting edge of science, particularly in physics. The normal parameters no longer hold.

Particles, or are they waves, appear and disappear, have bizarre, impossible characteristics, resonate simultaneously with others at the opposite side of the universe. There is talk of multiple extra dimensions beyond the three of space and one of time that we perceive, of gravity much weaker than theoretically it should be, of invisible matter and black holes. A rational world? Our forebears of the Enlightenment would have difficulty recognising it!

It seems that even here we are learning that we have to live with uncertainty. It takes a considerable maturity to do that, but perhaps that too can be a basis of faith.

18 Psychotherapy and spirituality

By profession I am a psychoanalytic psychotherapist, and my spiritual home is with the Quakers. In many ways I am very typical of both those communities, but in other ways I am not. Anything I may write comes out of my personal experience — but it is my experience, no one else's.

Being a psychoanalytical psychotherapist means that I was trained in the Freudian tradition — or more strictly post-Freudian, since the subject has changed and developed enormously in the sixty-odd years since Freud died. I belong to what is known as the Independent group, which I always feel is the professional equivalent of the Anglican church — it is very English, pragmatic, allowing a wide range of views but essentially well-earthed in biological realities. (That is my perspective. Others may differ.)

Freud, of course, was notoriously unsympathetic towards religion, and until quite recently it was not done to admit to spiritual leanings or a religious faith in my own professional circle. It was not something that was open to discussion. That has now changed, and colleagues are beginning to talk and write about spirituality.

Jungians, of course, have always been more open-minded on this issue, as was Jung himself. I suffer from the fact that I find Jung's writings unreadable. I have tried many times over the years to no avail. I have a blind spot here, and I have to accept it. What I have found is that as I get older I have come quite spontaneously to use certain key ideas of Jung, though I may have to translate them into different language.

It is important to see all great men, however great, as creatures of their time and culture. When Freud was forming his ideas, the fashionable intellectual ethos was Scientific Materialism, and he thought in those terms. It has been a long-lived fashion, but it is increasingly felt to be inadequate. We have become disillusioned with it.

Also, for Freud religion meant, I suspect, the Judaism of his family background which had only negative connotations for him. We must also remember that in Freud's Vienna, anti-Semitism was powerful and widespread, and a hindrance in his professional life. Again, more negativity. I do not think Freud had any deeper understanding of religion. He never came anywhere near knowing what it was all about. The spiritual dimension for him was no more than magical thinking, and not to be taken seriously. There are still those who think that way, but it is interesting to note that, with a few exceptions, at the present time they tend to be found amongst the humanities' community rather than the scientific one.

Psychotherapy and spirituality.: for me what underlies both these areas of concern is the search for understanding — to understand what makes people tick, then ultimately to try to understand what life is all about — a tall order!

A characteristic of human beings is that we have an innate drive to make sense of our world by discerning patterns in our sensory impressions. Indeed, we have a tendency to see patterns even where there are none. We put a shell to our ears and hear the roar of the sea (actually the blood pumping through our ears.) We hear voices in what is actually random, so-called white, noise which contains no information whatsoever. In times past, we had magical explanations for all kinds of things which we could not understand — hence the belief in witchcraft, and some curious religious ideas. It is as if any explanation is better than none — and indeed, individuals for whom nothing makes any sense become mad.

The future psychotherapist has, in my experience, a desperate need to make sense of other people, and this is a need that is acquired very early in life. I sometimes joke that we therapists acquire our vocation by the time we are two years old, and there is a lot of truth in that remark. Typically, we have a mother who is troubled. We want to help her so that she can give us the mothering we absolutely need at this early, dependent stage of our lives. We still have hope and faith that she can, and is willing to, be what we need; but we struggle to understand why things are not right.

Some small children, faced with a difficult beginning, just give up on people. They withdraw into a world of ideas, attempt to become self-sufficient, have poorly developed people skills. If they are blessed with enough intelligence, they can become scientists or academics. Other small children decide that people are unreliable, and pin their faith on managing the external material world. Some of these become engineers; others become compulsively acquisitive. Things are safer than people. Of course, there are innate differences between children from the beginning, and these play their part, but these earliest experiences are hugely significant.

The future psychotherapist has a compulsive agenda to understand, to attempt to heal, to put things right. There is an underlying assumption that, if properly understood, it does all make sense.

People ask for therapy for a variety of reasons — usually as a desperate last resort having tried everything else they can think of! They are hurting, they are stuck in what feels to be an insoluble dilemma, they cannot see their way forward. They are looking for answers, and expect the therapist to provide them. Part of our skill as therapists is to help them endure the pain and the muddle until it gradually begins to dawn on them that the answers actually lie within themselves, and that our joint task is to explore the emotional morass until some clarification begins to emerge. It can take a very long time, and it takes endurance and commitment because it is an uncomfortable and sometimes painful process. There are no magic wands.

I do not usually talk religion with patients unless that is a normal part of their vocabulary. If they have a religious faith, then it is legitimate to look at how it is used. Some use it creatively; others use it to defend themselves against possible insight and change, and the pain that goes with it. That I find difficult, and sometimes I have reluctantly decided to limit my attempts to intervene.

However, I have never worked with anyone over a long period of time — and here I am talking of years — without them eventually beginning to ask the ultimate questions — what is life all about; what is my life all about; why all this suffering; if there is a God, what sort of a God? They reach this point irrespective of their faith background, or lack of it. I do what I can to help them find a language that has meaning for them, and to find a way forward in what is a spiritual journey. As in their personal emotional explorations, I try to evoke their curiosity, to ask open-ended questions, to let them find answers that have meaning for them.

Back to meaning. To find meaning in our life is a requisite for our health and sanity, and it is an ongoing process. We have experiences; we try to make sense of them and then integrate that sense into our current perspective on the world.

Then, at some point, we have to accept that there are some things that are beyond our comprehension, and always will be. We literally do not have the capacity to encompass the ultimate meaning in the universe. We may touch its fringes, we may have some intuition which feels powerfully real to us, but that is as far as we can get. We are limited creatures, and it is a sign of maturity to accept that reality.

It is at this point that a faith, a religious tradition, can be helpful. Existential uncertainty is not easy to live with, and we vary in our capacity to tolerate it. What a religious tradition can give us is access to the accumulated wisdom and experience of others, past and present. However, it is only useful insofar as it is congruent with our own integrity. I always like the quotation from the Buddha: "Believe nothing, no matter where you read it, or who said it — even if I have said it — unless it agrees with your own reason and your own common sense." (563-483 B.C.)

I have so far written of the search for meaning. The other concept which unites the two subjects is that of personal growth and development. Just as we are born with certain innate propensities which lead us to change physically from children into adults, which govern our basic functions without any conscious decisions on our part, and which indeed, function best when we are least conscious of them, so we each have all manner of unconscious potential which motivates us and drives us during the course of our lives. You can call it temperament or aptitude or talent, but whatever name we give it, it demands fulfilment. The more we are able to develop these inborn abilities, the more contented we shall live; the more blocked and frustrated we find ourselves, the more discontented and miserable.

Now, the reality is that all of us have more potential than we can ever realise. Every time we make a choice in life, we inevitably turn our back on other possibilities. It is in the ongoing struggle to reconcile what we might be with the opportunities that are available to us that our personalities are formed, and we become ever more individual as people. It is this process which Jung was referring to when he talked about *individuation*.

Somewhere in the New Testament gospels, there is a quotation from Jesus where he admonishes his disciples to become perfect as he himself is perfect. I understand from biblical scholars that *perfect* is something of a mistranslation of the original Greek, and that that the word *whole* would better convey the meaning. This gives a different flavour to the task Jesus was setting his followers.

A theologian who speaks to my condition is the Anglican, Harry Williams, who was a Cambridge don and went on to become a monk at Mirfield. He wrote a number of books in the Sixties and Seventies, one of which he entitled *Becoming What I Am*. He believed that the aim of prayer, indeed the aim of the religious life, was to become what we had it in us to be, to become whole. If we did this in faith and trust, God could use what we were to His purposes whatever our limitations and imperfections. (Williams was a man who had had a breakdown and some personal therapy, so he was speaking out of experience.)

I have come to believe that just as we have an innate ability to acquire language, so we have an innate ability for spiritual experience and that this is one of the characteristics of being human. There were a lot of different hominids — proto-humans — before homo sapiens appeared on the evolutionary scene. Amongst the qualities that distinguished him from his forebears appears to have been the production of cave art; bereavement rituals, such as ways of burying the dead; and the use of symbols which suggest a belief in an after-life. These changes appeared suddenly in evolutionary terms, in what we call the *Paleolithic shift*. It was at much the same period that we acquired language. We had become a different kind of animal.

I suspect that the individual capacity for spiritual experience is variable. Some of us are more talented than others in that respect. I have heard it suggested that the truly religious never form more than about ten per cent of the total populace, even at times of widespread religious fervour. However, the research of the Alister Hardy Society suggests that most people, if asked in a non-judgmental manner, will admit to having experiences of the numinous — those times when our personal boundaries become tenuous, and we become aware of a sense of unity underlying all things. Some people have them rarely, some much more frequently — when listening to music, seeing something of overwhelming beauty, watching a sunset, at moments in a deep relationship. The person concerned does not always recognise these moments as having spiritual significance, but they are experienced as valuable — and treasured as such.

In our society, we have been reluctant to talk about them — they have not been scientifically respectable — so we tend to keep them to ourselves, not realising that they are part of our common heritage.

This innate capacity for spiritual experience — what Jung would call an archetype — for most of us emerges relatively late, in adolescence, alongside the capacity for abstract thinking, and for social concern. I have known a few people who seem to have it as children — and one thinks of Wordsworth's poem where he writes of heaven lying around us in our infancy. For others it seems to need the kind of event which shakes one to the core, which breaks through the normal structure of living, before it can happen. I am talking, I suppose of how open we are, or how defended.

How we interpret these experiences will depend on our religious background or lack of it. Is this, perhaps, an argument for giving children some kind of religious instruction? I have mixed feelings about R.I. believing it does as much harm as good unless very sensitively offered, but it does at least provide some words and concepts, and without those it can be difficult to think about one's experience.

A final issue. Spirituality has often been seen and talked about as another dimension into which we can escape from the mundane, everyday world, as a flight from the body, as something that is in opposition to our animal selves.

I find that a totally unacceptable perspective. In Christian terms it is a heresy — the Manichean heresy — with which the Christian Church has struggled from its beginning. In psychotherapeutic terms, it is a pathological development — a schizoid defence mechanism.[1]

The Manichean heresy is widespread, powerful and very seductive. One can see its appeal, particularly in times past when life was all too often brutal and painful. We find it in the many cults which were around in the Middle East at the time Christianity came into being; we find it in the Gnostic Gospels and in the beliefs of the Cathars; we see it in the late medieval period, in the flight into celibacy and monasticism; we see it in those anti-sexual attitudes which have spoilt so many people's lives.

---

[1] Schizoid means splitting. It results in a polarisation of the world into good and bad, black and white, right and wrong, body and spirit. It is a primitive kind of thinking which is normal in small children, but which hopefully we grow out of as we mature.

One can understand its appeal when life has become difficult and disappointing and unbearable. Understandable — but a mistake. The great mystics in all the major religions seem to me to have been very well-earthed people in their daily lives. So, in my limited experience are our contemporary monks and nuns. I deeply regret that the Roman Church imposed its ruling on celibacy on the indigenous Celtic Church of the British Isles, where many religious communities were mixed and included married couples with their children.

I actually think that the Manichean heresy is blasphemous. It assumes that the Good Lord got it wrong when he gave us bodies and everything that goes with being incarnate. St. Augustine of Hippo was a great man, but when it came to sexuality he was deeply flawed, and the Christian Church has suffered from what began as contamination by his earlier beliefs as a Manichean.

I see physical development and emotional development and spiritual development as a unity — as a path we all have to travel, simply because we are human.

William Penn, one of the first generation of Quakers, said: 'True godliness don't turn men out of the world but enables them to live better in it and excites their endeavours to mend it.' That is a very Quakerly statement. It is also, to my mind, a firm rejection of the Manichean heresy.

I am, I suppose, in theological terms, preaching an incarnation theology. It is not always very comfortable. It is as far as my understanding goes.

19 In the manner of Friends

I am a member of the Religious Society of Friends (Quakers), a persuasion which came into being in England in the middle of the seventeenth century. It was a time of great social, political and religious confrontation, leading to civil war.

Quakers, Friends as they call themselves, have no established creed and no doctrinal test for membership. All that is up to the individual conscience. They also have no separated clergy. It is a 'do it yourself' organisation, where all day-to-day business and administration is handled by the members.

On Sundays, or other days if the need arises, there is a *Meeting for Worship* held in the Meeting House. Friends gather at an agreed time and sit in silence, usually in an open square around a table on which there will be flowers. The meeting lasts for one hour.

Sometimes nothing is said at all, but such silent meetings are rare. Usually at a certain point, often twenty minutes or so into the meeting, some Friend will rise and speak — 'offer spoken ministry' — but rarely for more than fa few minutes. More silence will then follow before, perhaps, another Friend is also moved to speak. There is no limitation on who can speak, but a strong convention forbids ministry near when the meeting is due to end. After the hour Friends shake hands, the Meeting for Worship is over, and the Meeting Clerk rises to read any notices that there may be.

There are also other kinds of meetings, the most important of which is the Preparative Meeting, which settles all administrative arrangements and appointments of individuals to particular duties on behalf of the meeting. An unusual feature of this (and other) meetings where decisions are taken is that the meeting can procede only on the basis that all present must acquiesce. A minority of one is enough to block any proposal, as a result of which opposition is not entered into without a good deal of thought, knowing it to be the 'nuclear option'.

The Peace witness

Although Quakers have no settled doctrines, they do have some very deep-rooted traditions, one of which is the rejection of resort to war as a means of resolving human disputes. You do not have to be a total pacifist to be a Quaker, but it helps.

Recently I was talking with a woman of mature years whom I know well in some ways. She is what I think of as a natural Quaker — in her demeanour, values and attitudes — and indeed she is well involved with her Preparative Meeting, but as an Attender[1]. She said she had been asked if she would become a Member but felt that she could not since she was no pacifist. If it came to the point, she would fight.

Her remarks made me think of my own position. The 1939-45 war bred in me a fierce pacifism. Although only a child, the constant hearty propaganda of the Movietone

---

[1] Three sorts of people are seen at Friends' meetings. Enquirers are people who have come to see what Quakers are really like and what they do, but without commitment. Attenders are those who come to meeting fairly regularly, but have not yet taken the step of formally applying for membership. Members are those who have been through the simple procedure of applying for membership — hardly ever refused, except to those who seem not to understand what they are letting themselves in for.

News at the cinema (a regular weekly treat) sickened me. I could not ignore the reality that the people we were destroying, particularly the young airmen, were someone's sons, husbands, brothers, who were as significant to them as our men were to us[1].

One of the things I was to learn about myself subsequently was that I too could be capable of murderous thoughts and feelings. On the two occasions (both chronic, prolonged situations) when my life was in reality under threat, the urge to act out was very powerful. I managed to contain my feelings, but I gained some sympathy for those who could not. Moreover, I was never again certain that 'if push came to shove' I too would not fight.

However, this is not the point. The urge for self-preservation is a bit of innate fundamental biology which we all hold in common. Surely the point of the Quaker pacifist position is an understanding that violence is no answer. Violence only breeds more violence. We have to find a better way of resolving conflict.

Human beings are an aggressive species. This is why we have been so successful in evolutionary terms. However, it has also been our undoing. The civilisation of Classical Greece was finally destroyed by the states' continual warfare. My own ancestral tribe — the Welsh — were so intent on squabbling with each other that they could never get around to cooperating long enough to put up a proper defence against the Normans.[2] In the twentieth century the Soviet Union bled itself dry by spending too much of its income on armaments and too little on infra-structure; while the 1914-18 and 1939-45 wars devastated Europe,

---

[1] Naturally, at that time, we did not know of the horrors of the German concentration camps. These added another dimension to my young imagination, never to be forgotten.

[2] By the time of Owain Glyndwr it was too late in the day.

and we are still suffering from the long-term consequences of those conflicts. There has to be a better way. Our powers of destruction are too great for us to go on playing the old political games. If we do, whatever the rhetoric, we shall destroy all that matters to us.

A military culture

One result of living in a society always poised for war is the distortion of values that is engendered. The English do not think of themselves as a militaristic people, yet as a consequence of their history they are[1].

The Normans came as an army of conquest and occupation. They were a group of land-hungry robber barons who divided up the country between them under the nominal leadership of a king, who nevertheless had to continually manoeuvre in order to maintain his kingship. Over a period of centuries they settled down into a landed aristocracy who held their position by virtue of their function of raising men for an army which they officered, when called upon to do so.

Napoleon, with his concept of *levee-en-masse*, destroyed this time-honoured function of the aristocracy, and as a caste they have been in decline ever since. However, in England the legacy of this history is still with us in what we think of as our class system, and in our social organisations. Confrontation is a central part of our culture.

- Observe our obsession with football and other so-called games.
- Listen to the recordings of 'Yesterday in Parliament' on any day when the House is sitting.
- Compare our legal system with those of Scotland or France.

---

1 And are widely perceived as such by other nations.

Sport, Law and Parliament are based on what is a thinly-disguised and modified battle — it is war by another name. One can argue that these are harmless outlets for our innate aggression, and doubtless to a degree they are, but do the antics of our most senior politicians offer the best possible models for civilised adult behaviour?[1]

However, my major complaint about the distortions induced in a militaristic society is the effect on the role of women. Women have different values from men — complementary ones. These derive from their life experience of bearing and nurturing the next generation, but permeate all else. No one likes to see their life's efforts wantonly destroyed or treated as of little value. Yet this is what happens.

Soldiers cannot allow themselves to really feel what their womenfolk are going through in terms of anxiety, loneliness and the chronic fear of loss. They dare not look ahead to the after-effects that death and mutilation in battle will bring, because it would undermine their resolve. Their loyalty to their family has to take second place to their loyalty to their comrades, whose lives depend on them.

I am not belittling bravery. I have a lot of respect for the many service people I have known — but I have often grieved for their wives. Some of them found ways of surviving by adopting the same mores as their husbands, and many of them were brave and admirable people. It seems to help if you come from a service background and have had the education to match. I hope they felt it was all worthwhile. Joanna Trollope has a poignant portrait of one such wife in her novel *The Rector's Wife*.

---

[1] I am reminded that the Oxford English Dictionary gives a definition of the word 'debate', now obsolete but still current in the 16th century, as 'physical strife, fight, conflict'.

I have grieved even more for those children of the British Empire, sent 'home' to boarding school at much too early an age while their mothers chose to remain with their husbands. I also grieve for those children of servicemen who stayed with their parents. but who moved around the country so much that there was little continuity in their lives. They belonged nowhere, and they ceased to make friends because the recurrent loss of parting from them became too painful.

These are casualties of our system as much as the more obvious ones — the many single women left in the wake of both world wars, who subsequently lived admirable, unsung lives as best they could, but who missed out on so much through no fault of their own.

The rise of feminism

Is the basic (if covert) military culture of our society a factor in the rise of feminism in the twentieth century? It seems to me that women have rejected the role and values hitherto assigned them. They are no longer willing to live their lives in the service of an ethos which automatically relegates them to a second-class role and abuses their fundamental instincts of creating and nurturing life. If the massive loss of their menfolk in the two world wars has taught them anything, it is that they can live independent lives if they have to. They do not have to be mere accessories in the male game.

However, sadly, the price paid seems to be that contemporary women have come to value themselves only in male terms — a committed feminist is all too often a pseudo-male leading a masculine-type life, rather than a woman with a different agenda and role. Nowadays we have a situation where women expect to work both outside and inside the home, have both a career and bring up children; where the current cost of housing leaves too many of them little choice but to work to contribute to the joint income; where many are postponing maternity to an age far from optimal for mother or baby, and there is an increasing social problem of low fertility and expensive interventionist treatment We are then surprised that more and more women are developing male patterns of illness, and smoke too much. Children all too often are left in the care of people who do not love them, their emotional lives are impoverished and complicated from the beginning, with subsequent problems of behaviour and socialisation.

Happily, there are signs that those young women currently of university age are beginning to reject these impossible conditions, and re-think their priorities. Where they lead, others will follow, but we have created a difficult social ambience in which they must struggle.

Quaker pacifism

In the past, Quaker pacifism was couched in terms of the individual's resistance to state coercion. Quakers offered the state 'active obedience or passive suffering', and we honour the courage of those who were frequently treated very harshly as the result of their stance. In more recent times, much valuable work has been in the area of conflict resolution, and techniques for de-fusing potentially explosive situations[1].

We could usefully do some thinking about what gives both individuals and the wider society the capacity not to slip into fight mode—to restrain our aggressive and destructive impulses; to see the wider implications and to find ways of using our anger creatively. I am talking about emotional maturity, and this begins in the home, in the ordinary good-enough family.

I should like to hear Friends talking more in terms of the optimal conditions necessary for human growth and

---

[1] The creation of the Chair for Peace at the University of Bradford was a tremendous achievement.

development, at both the individual and social levels. In a sense we do address these matters in our social witness and the activities deriving from it. Historically our schools have done an excellent job; our treatment of the mentally ill has been ahead of its time; our social concerns and work for peace have defined us as a group. Intuitively, we have always sensed what is needed in order for people to develop their potential. What has changed is that now there is so much more knowledge available to support our approach. This knowledge ought increasingly to inform our understanding, and enable our thinking to move to another depth.

Aggression and the religious life

If a high level of innate aggression is a part of the human package, there are other aspects of our basic biology which have been long under-valued, over-looked and often silenced. In particular I have in mind the needs of the human infant and child, and the conditions which are necessary in the environment to allow those needs to be met. It is only when they are met that our society will have enough soundness to meet the challenges facing it, by virtue of the soundness of the individuals who comprise that society.

Ultimately, we have to ask ourselves what the religious life is all about. From an early stage in the history of Christianity, the message of Jesus of Nazareth has been contaminated by all sorts of ideas then current in the classical world, including what theologians call the Manichean heresy. This doctrine sees the spiritual and material as separate and opposed, religion being the domain of the spiritual, leaving the material world, including our bodies, despised. Perhaps it felt a comforting doctrine in those historical periods when life was all too often brutish, short and nigh unbearable. However, the mainstream Church has always rejected the doctrine, and been less than tolerant to those, like the Cathars, who would not relinquish it.

My own understanding of the gospel story — God who became Man — is as a statement about the Incarnation; that of God in every man, but also in all aspects of the created world. Such a perspective 'don't turn men out of the world but enables them to live better in it and excites their endeavours to mend it....' as William Penn so memorably put it. It also demands that we treat with respect that creation, and all that is in it, including ourselves. I resonate with the words of Alex Wildwood (2001) when he writes of 'the spiritual reality of the sacredness of the earth and the integrity of creation.' He goes on to ask: 'What would it mean to to be open to its ministries to us, to remember that we are not apart from but a part of the divine mystery of Creation?'

Karen Armstrong's long and difficult spiritual journey has finally led her (2011) to believe that the religious life is ultimately about becoming fully human. She says: 'I have discovered that the religious quest is not about discovering 'the truth' or 'the meaning of life", but about living as intensely as possible in the here and now.

The idea is not to latch on to some superhuman personality or to get to heaven, but to discover how to be fully human, hence the images of the perfect or enlightened man, or the deified human being. Archetypal figures such as Mohammed, the Buddha and Jesus become icons of fulfilled humanity. God or Nirvana is not an optional extra tacked on to our human nature. Men and women have a potential for the divine, and are not complete unless they realise it within themselves.'

Harry Williams, the Anglican theologian, said it even more simply in his small book on prayer (1977): 'For absolute love, God's love, makes us fully ourselves, instead of the half people we generally are. And to become fully yourself is a terrible risk. It would commit yourself to God knows what and lead you God knows where. If I open my heart in simplicity to God's love I might soon find myself in Bangladesh or something of that sort ... it might make me concerned about the oppressed people of the Third World or even about my neighbour next door who is lonely. And God's love has been known to make the most respectable people enjoy a pub crawl.' He had a nice sense of humour!

Williams believed that: 'In prayer we enter into the realm of reality and see things as they really are, from God's point of view.' I have always thought that seeing things as they really are is a strength of the Quaker tradition, and the Meeting for Worship provides us with a setting in which we can risk that exploration. I hope we shall continue that tradition in our ever more complex and difficult world.

Barriers: to break down or respect

Not long ago my own Quaker Preparative Meeting marked the occasion of Remembrance Day[1] in creative fashion. A group of Young Friends had spent the weekend with us, and on arrival for our Sunday gathering we found the archway between the meeting room and adjacent library blocked by a 'wall' composed of cardboard shoeboxes. On one side of each box were listed words denoting the world's woes; on the other side suggestions as to how those woes could be addressed.

Two extracts were read from a poem *Mending Wall* by Robert Frost, which had been distributed amongst us. The poet had been engaged with a neighbour in repairing a stone wall between their two properties. The neighbour kept reiterating the phrase 'Good walls make good neighbours' — a phrase beloved of his father. However, the poet queried whether there was actually any need for

---

1 Commemorating the two world wars of the 20th century.

a wall between them at all, since there was no livestock to contain and the vegetation made a clear boundary.

Friends were then invited to each retrieve a box from the pile and to meditate on the messages conveyed. This led to a lively and talkative hour, exploring all those issues of peace and justice which so preoccupy us in our social concerns, the need to break down barriers, build bridges and further our ideals in an active way. The symbolism of the 'wall' had been a powerful one. The young visitors did us proud.

However, I found my own thoughts not altogether consonant with the mainstream. I found myself looking at the walls which formed the building of our Meeting House and the much-valued space it contained, used by a great variety of people in the community as well as ourselves. I thought of how we all grow and develop before birth in an enclosed and bounded space in our mother's bodies. I thought of our homes, and how important it is to most of us to have a bounded space that we can call our own, and how we grieve for the homeless who have none. I thought of the social scientists who have discovered that communal space is so often trashed, while people who have their own front doors and small gardens are more likely to look after them with care.

I recalled the paper written by Winnicott, that sagacious paediatrician and psychoanalyst, entitled *The Meaning of the Word 'Democracy'* (1950). He suggested that the democratic process had developed to such maturity in Great Britain because we have had a natural boundary in the sea, as has Switzerland in its mountains, and that this process is more difficult to achieve where the frontiers between countries are arbitrary, and artificial barriers have to be maintained.

My imagination moved on from thinking about social groups and family groups to thinking about the individual cells which compose our bodies, and those unicellular organisms from which all living creatures have ultimately developed. It is a characteristic of such cells that each is surrounded by a membrane which separates its inner contents from the environment outside. This membrane is semi-permeable in that it allows some substances, but not all, to seep both into and out of the cell by a process known as osmosis. However, it is vital that this membrane remains intact if it is to perform its osmotic functions of primitive feeding and excretion. If it is damaged, the cell contents are extruded and the cell dies.

I am reminded that our human minds have something comparable to a semi-permeable membrane around each one of us. If it does not function properly, we are in trouble. Rigid 'wall-like' defences reduce our emotional flexibility and powers of adaptation. They inhibit our capacity to mature and develop. On the other hand, a failure of the psychological membrane can lead to overwhelming stimulation, and we are then in danger of becoming mad.

We manage the boundaries around our Meeting Houses with the skill born of long experience and communal goodwill. People, Friends and others, come in and out, all with their own purpose and agenda, making good use of the space and the facilities. We do well in this respect, and everyone benefits.

The healthiest families seem to function similarly. The family boundary is intact at any given time, but others — friends, visitors, new partners — are welcomed as they come and go. We rightly wonder about families who live enclosed lives which exclude others. They may have their reasons, but we worry as to what is going on behind the barriers.

I am searching for a metaphor, a perspective, which gives recognition to those innate characteristics which are part of our package as human animals, but which enable us to meet the challenges of our contemporary world. Those challenges — of a multi-racial, multi-faith society, of the current and massive world-wide movement of peoples experienced as immigration, of the incessant, unrelenting bombardment of communication leaving us unprotected from the miseries and conflicts of others — we cannot escape. The attempt to erect rigid barriers could only ever be a temporary solution, and within such barriers the organism begins to die. Yet to be totally open is to invite the rupture of the containing membrane, and another messy kind of death.

While I can feel with the rage and frustration and passion which underlies it, the well-intended image of 'breaking down barriers' makes me uneasy. It is too violent, and the resulting damage, individual and social, can be more than bargained for. I much prefer that other image of a semi-permeable membrane which allows movement in and out while maintaining intact that vital separateness and containment. In this analogous way we can continue to grow and change, as must all living things, while avoiding the splitting and rejection — us good/you bad — which is so destructive to individuals and communities.

20 Quaker mysticism

Both spirituality and religion long pre-date science. One of the great sources from which they derive is mysticism, itself surely the transitory perception of the ultimate reality that we can never fully understand — the breakthrough of the eternal into our three-dimensional, time-ordered world. At such times we make contact with the source of life, we touch the fringes of heaven.

I can talk only about my own experiences. They came thick and fast when I was an adolescent and young adult. Listening to music was powerful, as was contact with the natural world — the landscape, mountains, the open skies. Later came a sexual relationship, nursing my babies. At such times the world was in tune, a good place, a whole, and myself as part of that whole. This was in such marked contrast to my all-too-frequent mood of unhappiness, struggle and depression. These times carried me through. Without them I might not have survived.

It was only in my thirties that I came to make some connection with 'spirituality', and realised that these experiences were shared by others, both present and past. It rarely felt safe to talk about them, but I came to feel that within the Quaker context they would be understood.

Over the years, these episodes gradually diminished in frequency, though still tending to recur at times of stress. Now in old age, I rarely need them as once I did. They have gradually been replaced by a gentler, more continual sense of the world as God's ongoing creation. Like Brother

Lawrence, I feel I can make contact with God just as well at the kitchen sink as anywhere else.

Nonetheless, the shared experience of the Quaker Meeting for Worship almost always touches me at that deep level, and it remains powerfully moving (and rather exhausting!)

Did these experiences change me? They did not turn me into a saint; but they did enable me to change slowly from a tense, shy and rather withdrawn young woman into a reasonably sanguine and sociable normal adult, (and only those who have grown up feeling 'odd' know what an achievement it is to feel 'normal'!)

They did not send me out into the world to do great things, but helped me stay at home and attend to those matters that were under my nose and that only I could do, in spite of my frequent disinclination, boredom and frustration. I learned that creating anything of value is largely composed of such times of 'blood, toil, tears and sweat' and that it is from these times that comes satisfaction and joy and a sense of achievement.

Subsequently it gave me the courage to return to my professional training and qualify as a psychotherapist. As such, I put to good use the insights my personal struggles had given me, and I was able, so I have been told, to be helpful to a few of my fellow human beings in their own struggles towards growth and maturity.

Experiences of the numinous

I have come to accept the findings of the Alister Hardy Trust which suggests that these experiences of the numinous are universal in one form or another. Since we come out of eternity and finally return to it, is it surprising that we all have glimpses of it from time to time? Most of us do not talk about it because in our culture it is often suspect, even within the Christian church. It is not done!

And anyway we have no words. It is essentially an experience beyond words, though we may make the attempt to communicate.

The question remains for me as to why these experiences are more frequent, more available to some people than to others. It is as if for some individuals, at some times, the boundaries between this world and eternity become thinner, and there is a breakthrough. Wordsworth talked of heaven being around us in our infancy, and I have known a few people who had a powerful sense of God's love in their childhood. For some it occurs at the high points in their life, for others it comes at times of stress, of illness or bereavement.

There seems to be some link between religious experience and a tendency to psychosis, with some, such as George Fox[1], undergoing a temporary psychotic episode which they can subsequently integrate and experience as creative, while others go over the edge into madness.

There is a well-documented correlation between religious experience and epilepsy implicating a particular area of the brain — Paul of Tarsus being a famous example.

There is a long tradition among the religious of many faiths of fasting in an attempt to attain a sense of God's presence, suggesting that certain physiological changes facilitate it.

Since we are incarnate beings, it is hardly surprising if there are physiological accompaniments to these particular intense experiences as to any others. Such correlations seem to me to be largely irrelevant to the meaning and significance they have for people.

---

1 George Fox (1624-91) was a first-generation Quaker. He is generally regarded as the founder of the sect, and his experience of rejection in Lichfield when he saw the streets of the town as running in blood, ('Woe, woe unto this bloody city of Lichfield') would probably nowadays be understood as a temporary psychotic episode.

Certainly a too rigid clinging to the external material world, a habit of concrete thinking, appears to have a negative correlation. Perhaps one has to be able to let go of the 'certainties' in order to allow the 'unknown' to enter. Is this only possible if one has had the experience as an infant of falling apart and then finding oneself contained and upheld by the everlasting arms? If the human arms have not been there to rescue us from psychic disintegration, perhaps it is difficult, subsequently, to feel safe enough to let go.

Am I a Christian? I do not know. In the company of some professing Christians I feel comfortable and at home; with others I feel totally alien.

The traditions of ritual and doctrine have little attraction or meaning for me, though I can find them intellectually interesting. I can appreciate that they perform a function in encoding the ongoing experience of the Church down the centuries so that it does not get lost, and remains available to those who seek.

The exhortation and moralising of some groups I find positively off-putting, but I firmly believe with William Penn, 1693[1] that 'the humble, meek, merciful, just, pious and devout souls are everywhere of one religion, and when death hath taken off the mask they will know one another, though the diverse liveries they wear here makes them strangers.'

There are many paths up the mountain: we each have to find the way that is right for us. I am grateful for what I have been given, and for having found some good companions on the journey.

---

1 Penn (1644-1718) like Fox, was a first generation Quaker. They were men of very different temperaments, but each was impressive, and influential on those around him.

The Inner Light

Quakers are fortunate in their traditions — the emphasis on direct experience of the Divine, both individual and collective; the absence of a liturgy with reliance on spontaneous testimony; the lack of defining dogma, instead the collective wisdom found in *Faith and Practice*. The injunction to 'live adventurously' surely applies as much to our thinking as our doing.

All this facilitates a flexibility in our responses, a capacity for ongoing interpretation of what is needed in the Now; above all, an ability to trust in religion as continual revelation.

Listening to a Melvyn Bragg radio programme[1] on Soren Kierkegaard was fascinating. Here was a man who realised that *Reason* was not enough, not enough for a philosophy of life such as Hegel expounded. Each of us, he thought, has an inner life which is different, which demands to be lived and is outside reason. He understood that the world was such that there was no way our reason could encompass it. He saw life as a journey of faith, towards Faith; that we live in a world that is in process of becoming, and that that 'Becoming' is unknowable.

Teilhard de Chardin also saw the world as in the process of creation. The Anglican theologian H. A. Williams entitled one of his books *Becoming Who I Am*.

These ideas are far from new. *The Gospel of Thomas*[2], one of the earliest Christian writings, says:-

---

1 *In Our Time*. Radio 4. Thursday, 20.3.08

2 *Thomas* was rejected when Athanasius purged existing Christian scriptures in the 4th century, setting the boundaries of what became the New Testament. Writings not approved were destroyed wherever possible. *Thomas* was preserved by Gnostics, who probably redacted it to suit their own beliefs. Now seen as a Gnostic gospel, it might have started out mainstream Christian.

Logion 70 — Jesus said:
> When you bring forth that in yourselves,
> this which is yours will save you;
> if you do not have that in yourselves,
> this which is not yours in you will kill you.[1]

Logion 2 — Jesus said:
> Let him who seeks not cease from seeking until he finds;
> and when he finds, he will be disturbed;
> and when he is disturbed, he will marvel,
> and he shall reign over the All.

George Fox spoke of the Inner Light. He exhorted his fellow Quakers to open themselves to it, and that gradually it would reveal to them those hidden aspects of themselves, those inconsistencies, those little lies and self-deceptions which interfered with their inner peace and their capacity for authentic living.

The process sounds very familiar! It reminds me of my own personal experience of therapy and of being a professional therapist, the difference being that those journeys were made with a companion.

What we have inherited from our religious tradition is of immense value, all the more so given the cultural climate of our post-modern world. We should value it. We should take courage from it.

---

[1] A modern transliteration goes: When you bring forth that which is inherently within yourselves this which is yours will save you; if you do not acknowledge that which is within yourselves the alien introjects will kill you.

21 Life has a purpose

In the practice of psychotherapy the setting is all-important — the meeting with the same person in the same contained space at a regular time, creates a place of safety; and it is within this ambience that the thoughts and feelings of the patient's inner world can begin to emerge. Although some of Freud's thinking has been superseded by further theoretical developments in the hundred plus years since he published his seminal book *The Interpretation of Dreams*, his creation of the therapeutic setting remains a gift of inestimable value.

I see many parallels between this and the place of the Quaker Meeting in my spiritual life. The gathering in the same place at the same time with a group of similarly-minded people facilitates that turning of the inward eye to the depths — to begin to experience the universe as it actually is, to become aware of that powerful energy which created our universe, which continues to create and sustain it, and on which we are ultimately totally dependent.

My own scientific education left me with a profound sense of the order in the universe, of an ongoing ordering, indeed a creative drive, operating at all levels. It behoves us to respect it, to tune-in to it, to let it live us if we have the courage. It takes courage at times. 'It is a terrible thing to fall into the hands of the living God'.[1]

---

1 Hebrews 10.31

If I have learned anything in the spiritual dimension, it is that in the end all that is demanded of us is that we be fully ourselves, to be who we are and were meant to be. Our weaknesses and limitations can be used as effectively as our strengths and talents if we can allow it. What will be evoked may be surprising. What may be demanded of us maybe very big or very small. We rarely know the significance of what we do and what we are. This is where faith enters.

In his early professional life, Freud had a dream that neurology might one day provide a scientific basis and explanation of human behaviour. He came to accept that that would not happen in his lifetime — current knowledge and methods of investigation were simply not adequate to provide the necessary data. At this juncture he made the move from the study of neurology to that of psychoanalysis.

Psychoanalysis and the Bolsheviks

It is an irony of the 20th century that the attempt by Stalin to suppress psychoanalysis within the Soviet Union should ultimately lead to a methodology which seems likely to fulfil Freud's youthful dream.

After the 1917 Bolshevik Revolution in Russia, for a time many young intellectuals were drawn to psychoanalysis. Sabina Spielrein returned, after the ending of her relationship with Jung, to work with like-minded colleagues in a nursery inspired by psychoanalytical ideas[1]. Among her contemporary enthusiasts was a young neurologist called Alexander Luria.

However, when it became clear that Stalin was not going to tolerate this intellectual threat to Marxist ideology,

---

1 She soon fell out of favour with Stalin. Being a Jew, she was shot, with her two children, by the Nazi invaders (1942) in Rostov-on-Don, her birthplace. Her husband and brother had died in the 30s Stalin purges.

Luria opted for survival, retreated into his neurological work, and constructed a fruitful and successful career in that field. His research was detailed, thorough and scientifically sound, and out of it came many papers plus a book, published in 1977 shortly before his death, in which he expounded an innovative perspective on brain organisation which he called dynamic localisation.

His work demonstrates that specific psychological functions are not located in specific areas of the brain as was once thought, but that these areas contribute their particular functions to a systemic organisation of which they are an essential part. In other words, the psychological function is located dynamically between a number of specific areas of the brain fabric.

A book by Karen Kaplan-Solms and Mark Solms addresses this perspective, exemplifies it convincingly with clinical material and sees it as the way that psychoanalysis can come to have a sound neurological basis. (Kaplan-Solms 2000) Indeed they make a convincing case for the dynamic location of some of Freud's earliest theories. Work is ongoing, there is more confirmatory material than is yet published, but vistas have opened up in a fascinating way. We are only at the beginning of an intellectual journey which will hopefully integrate several hitherto diverse strands of conceptualisation.

Having said that, this is scientific enquiry. As such it may tell us much about how we function. It can tell us little or nothing about why, what motivates us, what make us who we are as individual human beings.

Semir Zeki is a visual neurobiologist[1]. His book *Splendours and Miseries of the Brain* also addresses the neurology of the brain, but from a somewhat different perspective. (Zeki 2009) His scientific background is as impressive as that of

---

1 Department of Cognitive Neurology at University College London

the Solms, but he lacks the psychoanalytic perspective that they bring. Yet he too understands that the brain — the physical entity — structures, both literally and metaphorically, our psychology — the way we think and behave.

He sees certain tendencies of the human condition as innate, hard-wired[1]. He talks of inherited brain concepts or programmes — the innate organising capacities of the brain which allow us, for example, to see colour, to recognise objects from whichever angle we view them (form perception), to conceive of 'unity-in-love'. These primary capacities allow us to organise our perceptual experience, and to then develop abstract concepts which facilitates the acquisition of knowledge. They are however, surprisingly specific, and a number of different primary concepts may be involved in an apparently simple human attribute or thought or perception, each concept localised in a different but specific part of the anatomical brain.

These primary programmes have certain qualities, Zeki argues. Firstly, we cannot discard, ignore or disobey them. Secondly, they do not change with time or experience — they are immutable. Thirdly, the systems of the brain concerned with these capacities are relatively autonomous of other such systems.

Acquired brain concepts are distinct from inherited ones in that they are developed by the brain throughout its post-natal life, and are therefore capable of being continually modified. Their development depends upon the acquisition of new experience which is then incorporated into, and integrated with, earlier experience. Memory and judgment are significant factors in what is acquired, so many areas of the brain are involved in their

---

[1] They are somewhat analogous to the operating system of a computer, though immensely more complex.

development — they are synthetic concepts. They differ from primary concepts in that they are subject to an ongoing process of change as they undergo modification with experience.

It is Zeki's background in neurology, in particular his work on colour vision, that has led him to the ideas he expounds in this book. It enables him to exemplify his theories in great and convincing detail. While he, too, is interested in how the brain functions, he is more concerned with the psychological meaning its function has for us, as we attempt to live our lives. He brings to the subject a wide cultural background, from the philosophical (Plato and Kant) to the literary (Dante), from music (Wagner's opera *Tristan and Isolde* is a major topic), to art (Michelangelo and Cezanne), from the *Song of Solomon* to the Sufi mysticism of Ibn al Arabi.

However it is when he departs from his professional expertise that I find him less convincing. In particular, he lacks the background in psychoanalytic thinking that makes the Solms' work so interesting. Though thoughtful and stimulating as he interweaves his various threads, Zeki leaves me uneasy with some of his basic assumptions.

He sees the major driving forces in our lives as the drive to accumulate knowledge, and the drive to achieve union with the loved one. Of the former he says, 'The splendour of the brain is that it is capable, seemingly effortlessly, of generating so many concepts and thus acting as a very efficient knowledge-acquiring or, if one prefers, knowledge-generating system. However, the misery that this splendid machinery brings us is in fact the result of its very efficiency. The incapacity of our daily experience to live up to and satisfy the synthetic concepts that the brain generates, commonly results in a state of permanent dissatisfaction.'

Beyond the pleasure principle

It is many years now since Freud expounded on the tension between the Pleasure Principle and the Reality Principle, which is what I sense Zeki is commenting on. (Freud 1920) It is surely an inevitable, and indeed necessary, part of our developments as human beings that we struggle with this tension. The achievements of our lives emerge from our attempts to integrate this tension — in analytic terms, to achieve the depressive position. I have no sense that Zeki understands this.

I am also uneasy about his idea of 'unity-in-love' as a primary concept. The notion that it can only be achieved in death was a powerful Romantic fantasy, and he waxes eloquent about it as portrayed in Wagner's *Tristan and Isolde*. Again, such a position is surely an expression of the Pleasure Principle over the Reality Principle.

We all have an experience of 'unity-in-love' at the beginning of our lives, in the womb. Any of us are capable of regressing at times of stress to wishing we could retreat back to that state. I have tended to see it as an expression, in Kleinian terms, of the Death Instinct — a turning away from life, as essentially, if only temporarily, pathological. To live one's life in that hope, as that ultimate aspiration, seems to me highly suspect.

In reality, where such relationships exist, where one partner totally subsumes his or her personality in the other, the result becomes exceedingly boring; and the dominant partner frequently breaks out by finding outside stimulation in a sexual affair or some other engrossing activity. Even where it is a true *folie á deux*, a joint creation, there is no room for children, for new life, Any children of such a relationship feel excluded and neglected. It is a non-creative sterile unity.

Seen as the ultimate goal of the religious life, I can give it a worthwhile meaning; but even here it leaves aside the question of what our life is all about. For what purpose were we put on this earth? To see life as an essentially miserable, if occasionally glorious, episode between the womb and ultimate fusion with the Divine, devalues the whole of Creation. Indeed it smacks of the Manichean heresy, the material world seen as evil.

Although as we go about our daily tasks we feel we are living our lives, there is a sense in which Life is living us. It is our inborn programming which ensures that, given a good-enough environment, we grow from babies to children, learn the skills of bi-pedal locomotion, learn to talk[1], traverse that singularly human phase of socialisation we call latency, then move into sexual maturity with all that that involves. It is our innate biological imperatives which sets the agenda; we can only influence the forms it takes.

This perspective equates with Zeki's notion of primary concepts. This is the point to which evolution has brought us.

At the same time, we each of us have to struggle with who and what we are. In addition to that innate potential which we share with all other human beings, each of us has our own personal individual set of abilities and talents which demand expression.. Some of us are fortunate to recognise these, or receive recognition and encouragement from others at a crucial time. We develop interests, a vocation; and the need to develop these drives us inexorably and shapes our future lives.

For some, the need to have children is paramount. The desire to create a family is all they ever want, and that

---

[1] What particular language we acquire depends on what is spoken by those among whom we live. Even in that sphere, though, there is a limited window of opportunity. The few rare examples of feral-reared children suggest that if they have not had the opportunity to learn speech by a certain age, they no longer have the capacity to acquire it.

creative urge should be respected as perhaps the most important of all our human achievements. Others are driven to become artists, musicians, actors, scientists irrespective of whether the decisions they make are sensible or reasonable.

People who follow their star rarely regret their lives, whatever the hardship. By contrast, those who for whatever reason find their natural path of development blocked or unavailable can reach the end of their life in disillusion, with a sense of waste. It is as if our innate potential demands instantiation. I am trying to describe the process which Jung refers to as individuation. It is about becoming who we uniquely are — a lifetime's task.

At the same time as we are being driven by the incessant demands of our human and personal nature, we have a more conscious task of trying to make sense of it all — of ourselves and the world we live in. Zeki sees the brain as programmed to acquire knowledge. I see it more as programmed to search for the meaning of it all.

It is an ongoing process, beginning at birth and continuing as our world expands. It is a process of sifting and then integrating information into a satisfying whole, which goes on while we are both awake and asleep. It is never ending until death. Crucially, it is dependent in its beginnings on having a mother who brings the world to her baby in a way that allows it to make sense of its initially very limited experience — not too much stimulation, not too little. If, sadly that mother is unpredictable, chaotic, so that no sense can be made by the baby, there is a real likelihood of that person giving up the attempt. The end result can be a turning away from people into a schizoid world, a turning towards the reliability of things rather than people, at worse a retreat into the madness of schizophrenia — the abandonment and destruction of meaning.

That search for meaning can take many forms. It drives scientific research, artistic endeavour, philosophy, psychology. Ultimately it drives the search for God — the Ultimate Meaning. The reality that we know we never can know God does not inhibit the search. To search is the way we are made. This need to search makes us what we become, underpins all our human achievements.

Both the two Solms and Zeki write about the search for further knowledge, for further understanding and meaning. Advancing techniques of brain scanning have opened up new horizons in our comprehension of brain function. We are beginning to feel that before too long we may be able to integrate the various disciplines of mental life into some kind of unitary perspective.

Each of the two books is a brave attempt along that path and worthy of attention. Each brings a perspective that needs to be addressed. I am filled with a wry amusement that, after so many years in the academic wilderness, Freud is being reinstated as an intuitively accurate pioneer in the field of neuroscience.

22 Religion in a post-modern world

There is a useful distinction to be made between *Spirituality* and *Religion*. I define *Spirituality* as an awareness of, seeking for, yearning for

- something greater than ourselves
- something which makes sense of our lives and our world
- the ultimate, the Divine

We all have the innate capacity for spiritual experience just as we have the innate capacity to learn to talk. It is something we are born with, part of being human. Some people have a greater capacity than others — it is a matter of temperament. It can be cultivated, or it can be neglected and ignored.

For most people spirituality seems to develop in adolescence at much the same time as the capacity for abstract thinking. A few seem to have it as children; others need an intense experience as adults — falling in love, a bereavement, a shock, an unexpected jolt out of their normal rut — in order to become aware of this dimension.

It is not something we easily talk about in our culture, but most people if asked will confess to moments in their life when they have had an experience of the numinous, which has made everything feel worthwhile. It may be triggered by watching a sunset, by listening to music, by a moment in an intimate relationship. It comes out of the blue; it is timeless. 'Prayer is the world in tune' wrote

Henry Vaughan, the seventeenth century poet, in an attempt to capture the essence of such an experience.

So — *Spirituality* is always with us, independent of time, culture, fashion. The most gifted of spiritual human beings can speak to us down the ages.

Religion I understand as the organised social expression of this innate fundamental yearning. By contrast, it varies widely with the time, the place, the culture and the temperament of the group. All religions tend to develop rituals, liturgy, dogma, sacred places, sacred buildings and their own specific art forms . Religions are inclined to get involved in the politics of their day, and never fail to have their own internal power structures and tensions. They all tend to assert that they, and they only, possess 'the ultimate truth.'

I want to suggest that our religion is an inextricable aspect of the particular world we live in. The classical world of the Greeks and Romans was arbitrary, unsafe, fragmented and incoherent — and their gods reflected just those qualities. The emergence in Christianity of a monotheistic, consistent world picture must have been a revelation. It also had the added bonus not only of an acceptance of pain and suffering as part of the human condition, but also hope — hope that it could be survived and, above all, given meaning. No wonder Christianity swept the Mediterranean world like wildfire. Here was a picture of God which could make sense of our world and our lives.

The Roman Empire was run on a political system of bureaucratic centralism. It only ever worked well during the empire's expansionist phase. When this was thwarted — by the growing strength of the barbarian peoples and the ever-lengthening lines of food supply — the empire retreated progressively. It lasted a long time. It gradually became what would later be called the Byzantine Empire, centred at Constantinople, with its Eastern Church using the Orthodox Catholic rite. After many successful centuries, during which it spread its own flavour of Christianity to Russia, the Balkans and beyond, it finally disappeared from view before the Muslim advance in 1453.

At an early date the Church in Rome made a bid for the primacy of the five Christian patriarchies of Alexandria, Antioch, Constantinople, Jerusalem and Rome[1]. It was heavily influenced by the Germanic tribes who had over-run most of the Europe formerly held by the Roman empire, and developed a military-style organisation which accommodated it to that Germanic culture. Power was all, and the fact that the other four patriarchies expelled the Roman Church was ignored and denied as it proclaimed itself the One True Church. The saving of souls was seen as the most important work of the Church and to that end anything was justified — a good story was preferred to historical fact, improbable saints abounded, documents were forged[2], and alternative myths about the early Fathers were propagated.

The darker aspects of subsequent church history — the attempts at conversion by intimidation, sword and fire — were inevitable consequences of this initial insistence that there was only one true church. It is a marvel — a miracle — that the central insight and teaching of Jesus of Nazareth was grasped and practised by so many good people in spite of the church hierarchy and its deviations.

---

1 The struggle for power in the Church began in earnest at the Council of Chalcedon in 451, continuing until finally settled in 1054 with mutual rejection by Constantinople and Rome — 'the Great Schism' — after which there was no longer a single Catholic Church.

2 The *Donation of Constantine* was an 8th century document purporting to be from the 4th C. The *False Decretals of Pseudo-Isidor* were a similar forgery from the 9th C. Both were intended to forward the claims of the Bishop of Rome to be Primate of the Catholic Church.

If this judgment on the Roman church appears harsh, it was not the only Christian group to manifest discrepancies between its overtly expressed ideals, and much appalling behaviour committed in the service of those ideals. However, such behaviour, when examined, is usually typical of the culture of a particular historical period rather than anything derived from basic Christian principles — though Judaic scripture may sometimes be quoted in defence. The reality is that it is only exceptional people who can think and act outside the norms of their time. Fortunately there are always a few such.

The worst excesses of misplaced religious zeal occur when the Church plays politics, since politics are by their very nature short-term and of their immediate time. An attempt to serve both masters is nearly always to the detriment of true religion. Jesus had something to say on this point![1]

Religion characteristically maintains that it is concerned only with eternal truths. In reality it is always adapting to the attitudes and ideas of the given culture and period of history in which it operates. That is part of what it is about. Each generation is faced with challenges, and a monotheistic religion has to somehow meet those challenges and integrate its responses into the main body of faith.

Christianity has been such a successful religion because it has proved to be remarkably adaptable to different peoples and cultures, whether by schism or evolution.

What are the challenges of our time?

For us, a viable religion must be compatible with our scientific understanding of the universe. While the science of the 19th century, with its emphasis on materialism and reductionism, appeared to stand in opposition to religious

---

[1] 'Render unto Caesar the things which are Caesar's, unto God the things that are God's.' Matthew 22.21

faith, more recent developments in scientific thinking have produced a different mindset.

The more we explore the boundaries of our knowledge, the more certainty crumbles. The world begins to look very much odder than we ever imagined. Scientists are now talking in terms of a hyper-dimensional universe of which ours is just a small corner! It is increasingly accepted in the scientific community that science is not about 'Truth' but only about models of the outside world which work more or less well, and which have to be continually updated. We are beginning to suspect that ultimately it may all be beyond our comprehension — a salutary experience.

This acceptance that perhaps we can never know or understand the whole truth may help us come to terms with what I see as the other great challenge of our time — the impingement upon the Christian and post-Christian world of other belief systems, particularly that of Islam. Communication of every kind is nowadays such that we can no longer live within our own bubble of belief, and ignore what is outside it. We have to somehow learn to accommodate the different perceptions of God, the different interpretations of what constitutes the good life, and the different beliefs about the truly religious response to the problems of being in this world.

This is in stark contrast to the fundamentalism which appears to be flourishing in our time — and not only in the religious domain. I see it as an attempt to stake out territory which feels safe, at a time when the world feels very unsafe. It is a response to profound existential anxiety.

Karen Armstrong has pointed out how it is only in recent times, since the nineteenth century, that people have begun to interpret the Bible literally. In the past, it was certainly studied intensively, but in a way that sought to evoke understanding, to help the student intuit more of the nature of God, as encoding signposts on the spiritual journey.

There is a vital paradox operating here. The more one insists on a literal interpretation of scripture, Christian, Muslim, Marxist or whatever, in an attempt to feel safe in a frightening world, the more unsafe one actually becomes. Where people feel that they, and they only, have the truth, others are going to fight them. Such an attitude invites conflict; it sets it up as inevitable.

If we can accept that none of us has the whole truth, that for all of us our understanding is partial, incomplete, then we can find a way to live alongside each other and be enriched by our different perceptions. It demands humility, the abandonment of entrenched positions, and the recognition of our limitations.

If asked whether I believe that God speaks to us in the Scriptures, I would answer 'yes'; but that our reception of the message is distorted by the limitations of who we are, by our preconceptions and unconscious assumptions, by the language we use when we try to think the unthinkable, and communicate the incommunicable. This is true for us now; it was true for all those other devout people who have left us records of their thinking and actions in the religious realm. All should be approached with respect — and a healthy scepticism. We need to hang on to the indisputable knowledge that we, all of us, have inevitably got it wrong. If we appear to be in conflict with equally sincere and devout others, it is because we are incapable of comprehending the Ultimate, the All.

That should not prevent us from trying to further our understanding, just from assuming we have finally got there. Quakers are fortunate in their traditions — the emphasis on direct experience of the Divine, both individual and collective; the absence of a liturgy with reliance on spontaneous testimony; the lack of defining dogma, the assumption that we are all seekers after truth.

All this facilitates a flexibility in our responses, a capacity for ongoing interpretation of what is needed in the 'Now'; above all, an ability to trust in religion as continual revelation.

What we have inherited is of immense value, all the more so given the cultural climate of our post-modern world. We should value it. We should take courage from it.

23 Credo

What do we have in common? We are all human animals, and we share the basic biology of the human animal. Many religions have gone off the rails by trying to deny these realities — by attempting to suppress what is irrepressible, by retreating into 'other-worldliness'. These impositions distort people's lives, and ultimately break down. What is basic to humanity?

- We are sexual beings; we are aggressive in defending our territory; our young are very slow to develop to adulthood and need a long period of dependent nurturing.
- We are social animals. We are designed to live in family groups, and in larger but still essentially face-to face groups. [1]
- We have an innate need to make sense of our experience, and of the world we inhabit. If we cannot, we experience existential anxiety which rapidly becomes unbearable. So if we cannot make obvious sense, we make up stories to explain matters to ourselves. Any story is better than none.
- We are born with an innate religious[2] sense. This may evoke a response for greater closeness to the

---

[1] Beyond this, power struggles develop — and conflict — and politics.

[2] By religious, I mean a sense of something greater than ourselves, which holds our world together and on which we are dependent.

> source of our being, greater comprehension, awe, fear, comfort, love.

These last two factors are the basis of the world's many religions. The stories people make up vary with time, place, culture, sophistication. They are all attempts to comprehend the incomprehensible. They are all inadequate. Ultimately what they seek to grasp is beyond our understanding.

From time to time there are a few remarkable people who see the world as it really is. They have an intuitive understanding and experience of the numinous which illuminates their lives. It is 'the pearl of great price'. It is beyond words.

They attempt to communicate their understanding to others, but the words are always inadequate. They can only ever be symbolic, poetic, allusive. Unfortunately the words are all too often taken by their associates as expressing concrete reality.

Stories are created, rituals developed, patterns and organisations set up, on the basis of an inadequate comprehension. A religion is formed. It is comforting, it is illuminating; it is the best the group can manage.

The trouble comes when they come to think that they have 'The Truth' and the whole truth, and that everyone else is wrong. Sadly, they cling to this conviction because their religion gives them comfort. It protects them from that existential anxiety which is so unbearable. They cannot afford to have it challenged.

The only way out of this dilemma, and the tensions arising from the clash of religious ideologies, is to face the fact that none of us can know the whole truth. It is beyond us; it is mind-blowing. Our brains are not adequate to grasp it. We have to learn to bear that anxiety.

We need to respect each other's attempts to struggle with Ultimate Reality and perhaps learn from those alien attempts. We should recognise and respect that that was the best people could formulate at that juncture. Our reality is that we are all of us trapped in the web of family, trapped in the web of culture, trapped in the web of Time. That is the meaning of Incarnation. We can only live and work within those constraints.

I believe, as did Teilhard de Chardin (1965), that we live in a world that is in the process of creation, and that we are a part of that process, for good and ill. We each have a part to play in it. We can only make our contribution by being who we are to our fullest potential. We cannot be other than ourselves. We should not live so as to satisfy others' models and expectations, and it is blasphemous to try.

Our contribution is always less than we hope, but more than we ever know. Our influence, often through small casual contacts, can be formative — creative or destructive. Rabbi Lionel Blue writes of angels unaware — the Jewish concept of angels as messengers from God, in ordinary human form. We all have that potential in our daily contacts.

Karen Armstrong, who has herself made a remarkable spiritual journey, and who has in the process investigated and written extensively about the various major religions, has come to the conclusion (2010) that at the heart of all of them is a profound compassion. That is what drives the universe. That is what we need to access. That is what can unite us.

We all have a developmental task inherent in being human, to become what we have it in us to be. The pattern of the task is one we have in common by virtue of our humanity; the detail is unique to each of us. In embracing that task, we involve ourselves in the process of creation.

The value of our individual contribution, we can never know.

By offering ourselves up to that process, we may find ourselves in strange places, doing unlikely things. It takes a kind of courage. 'It is a terrible thing to fall into the hands of the living God!' Yet it is our only salvation.

24 The journey toward P6

The capacity to experience the numinous seems to be innate in human beings — a neuro-psychological 'given'. It is more highly developed in some than in others. There are those who do not even recognise it for what it is. Only by sensitive questioning can one evoke their memories of such experience.

The words people use when attempting to describe something so intangible are various. One of the functions of religion is to provide a language in which to talk about it, and a framework within which to develop one's search and understanding of this tenuous but often hugely significant aspect of life.

Religion helps us to establish our social identity. By aligning ourselves with a given group of people, we assert our place in the social order and in the historical perspective. We signal to others something significant about who we are.

Religion helps us, as individuals and as groups, to cope with existential anxiety. It offers us an intellectual and emotional framework, 'certainties', which keeps at bay the fear of the unknown and the paralysing anxiety that we might be overwhelmed by the vast and terrible world in which we live. It helps protect our vulnerable inner core.

Where religion has been banned for ideological reasons, it is replaced by a creed which offers the same protection, the same kind of certainties. Communism is a recent example.

The more anxious people are, the more rigidly they cling to their 'certainties', and the more vociferously and actively they defend them. This is why there has been an upsurge of religious fundamentalism in recent years, notably in Islam and in the USA. It is, needless to say, a positive feedback situation. Each group is made ever more anxious by the certainties of the other, and so becomes progressively more rigid and belligerent.

Living with existential anxiety, with 'not-knowing', can be uncomfortable. It requires considerable emotional maturity. It needs a stable sense of self and self-worth which is ideally given by a warm and secure infancy and childhood. Deprived of this, it is difficult to achieve as an adult, though one can aspire to it, and work towards it.

Our world needs a religious formulation which accepts that ultimately we cannot 'know'; that all religions are an attempt to get at the truth, but are inevitable limited and failed attempts; that we can learn from each other on our spiritual journey, but that no one has all the answers. The ultimate truth is beyond us, beyond our capacity to grasp, to understand. In accepting this we can worship together, celebrate our differences, talk rather than fight, search for the common insights, get to the experience behind the words, grow as human beings.

Any religion for our time must be congruent with our current scientific understanding of the world, and also with the psychodynamic perspective of human development. Implicit in this is an acceptance of the material world as God's creation, and a rejection of the Manichean heresy which splits the spiritual domain off into a separate category. The denigration of ordinary life which this heresy has historically promulgated is destructive of human health and happiness. It has resulted in the most appalling behaviour towards others, such as burning heretics, and an abuse of our shared environment which now threatens us all.

All the major religious revelations have occurred at times of great social stress, frequently war situations. It is at such times, as in the personal domain, that the 'normal' barriers become thinned, and sensitive individuals become acutely aware of the eternal dimension.

Our society is entering such a period of disorganisation and breakdown at present. The current economic mess seems characteristic, as is the erosion of professional and public morals — the recent financial misdemeaners at Westminster, motor insurance companies routinely trying to steal each other's business; the 2011 scandal around phone-hacking in the service of newspapers, with the suspicion of associated widespread corruption, these are symptomatic.

As the Eurozone struggles with its internal tensions, one could well argue that its very creation was symptomatic of perverse thinking. At the heart of it is a denial of the profound cultural and historical differences between the various nations comprising this artificial construction. To imagine that these would disappear at the stroke of a bureaucrat's pen is to rate fantasy over reality. Also amongst those clamouring for ever larger bail-outs to rescue the dire results of financial incompetence, there would seem to be the assumption of inexhaustible supplies upon which to draw. At some point, the answer has to be 'no'.

Meanwhile, in a wider perspective, the inevitable march of climate change is going to disturb all sorts of precarious balances. The current large scale migration of people from poorer to richer areas of the world is likely to increase, as is the resistance to it.

It is surely out of this chaos that Hans Küng's *P6*[1] will be born. How soon? Who knows?

---

1 Short for Paradigm 6, which Küng sees as the next developmental phase of Christianity. See Küng (1990)

25 Riding the waves

Much of my professional life has been concerned with individual pathology. In this book I have attempted to address some issues that I see as social pathology.

What they have in common is denial and neglect of the essence of who we are; plus the attempt to control things that are beyond our control out of fear masking as arrogance.

We are mammals - highly developed, complex, thinking creatures - but still mammals, endowed with all those innate drives and instincts necessary to ensure our survival and reproduction. We are sexual beings. Our life patterns are driven by this fact; our art, our literature, is infused and obsessed by it. Our attempts to suppress or deny this reality damages both us and those around us. The Roman Church is currently grappling with this issue.

We are one part of the natural world, and ultimately dependent on that wider natural world and its integrity. If we damage that too drastically we shall perish.

We have a remarkable ability to think, to make tools, to manage and modify our environment - but ultimately we are still dependent upon it.

Our capacity for thought has given us much insight into how our world functions, and brought many blessings. However, it has tended to mislead us into thinking we can control it - and we can't.

We have over-valued our capacity for thought at the expense of ignoring those deep forces which drive us. We have done this at the individual level, but also at the social level. Our wider society - local, national, international (and familial!) - is driven by powerful primitive forces which shape our human community. Our artists reflect these forces back to us; our politicians are sensitive to them and respond accordingly. One of the weaknesses of democracy is that politicians who are not sensitive and responsive do not get elected. The result is that our leaders too often collude instead of leading.

It is some of these unconscious social forces which I have been trying to address in this book. It is really the realm of the anthropologist - preferably one with a psychoanalytical background![1]

One of the results of our infatuation with our own thinking processes and cleverness is that we have tended to split them off from our feeling and intuitive capacities, including our sexual drives. Organised Christianity has been particularly prone to this error. Some religious formulations are not as guilty - Judaism, Islam and Buddhism seem to me to be healthier in this respect, though they too have their less sound enclaves.

The emotion-free, supremely rational Mr. Spock, of the television series *Star Trek,* has been promoted as an icon to emulate. In reality, human beings who, through accident, have lost their capacity to have feelings are severely crippled. They are unable to think creatively or indeed to

---

[1] Wilfred Bion (1897-1979) made some significant contributions to our understanding of the unconscious forces at work in groups.

make decisions of any kind. It would seem that our emotions are the vital under-pinning of our thinking life. (Damasio.)

In the psychoanalytical world we have learned that the more aware we can be, as individuals, of our unconscious feelings and motivations, the more comfortably we can live with them and manage them. We learn to control through use; we learn what can be modified, and what has to be accepted and lived with. This is an important part of the therapeutic process.

We need to develop a similar process for those primitive and largely unconscious processes at work in our societies.

We are too prone to behave like King Canute[1] in his chair at the edge of the sea, instructing it to retreat. Instead we need to learn the skills of the surfer on his board and to ride the tidal waves. In doing so we shall undoubtedly experience some tumbles; we shall need to learn to swim so that we can clamber back to safety, but we might also find the same joy and exhilaration in going with the irresistible vitality of the surge.

Above all, because it is our natural element. we need to have the faith and the courage that it will carry us to shore, and to foster that faith and courage in our children.

---

1 Canute (Knut) - d 1035 - did not expect the sea to obey! He was making the point to his courtiers that he did not have unlimited power.

References

Ardrey, Robert (1970) *The Social Contract* (Fontana)

Armstrong, Karen

(2005) *The Spiral Staircase* (Harper Perennial)

(2007) *The Bible: A Biography* (Atlantic Books)

(2011) *Twelve Steps to a Compassionate Life* (Bodley Head)

Bion, W.R. (1961) *Experiences in Groups.* (Tavistock)

Blue, Lionel (1985) *A Backdoor to Heaven.* (Fount: HarperCollins)

Chasseguet-Smirgel, J. (1985) *Creativity & Perversion.* (Free Association Press)

Clerk, Rotraud de (2010) *Letters from Berlin: Alix Strachey's view of pre-fascist Germany 1924-25* in *The British Journal of Psychotherapy vol. 26, no.4.* (Wiley-Blackwell)

Damasio, A. (1994) *Descartes' Error: Emotion, Reason & the Human Brain.* (Putnam)

Erikson, Erik H. (1950) *Childhood & Society.*

Gosling, William (2001) *The Definite Maybe* (Xlibris)

Isherwood, Christopher, (1935) *Mr. Norris Changes Trains.* (1939) *Good-bye To Berlin.* (Vintage Classics)

Kaplan-Solms, K and Solms, M. (2000) *Clinical Studies in Neuropsychoanalysis* (Karnac)

Klein, Melanie (1935) *A contribution to the psychogenesis of manic-depressive states.* (Int. J. Psycho-Anal. 16: 145-74)

Kuhn, Thomas S. (1970) *The Structure of Scientific Revolutions* 2nd edition (Univ. of Chicago Press)

Küng, Hans (1990) *Theology for the Third Millennium: An Ecumenical View* (Anchor Books)

Leigh, Rosalind (1979) *'Panel Report on Perversion'* from the Barcelona Congress 1979 of the IPA. IJPA vol. 79, pt. 6.

Lamont Brown R. (2005) *Alice Keppel & Agnes Keyser* (Sutton Publishing)

Long, Susan (2008) *The Perverse Organisation and its Deadly Sins.* (Karnac)

Ogden, T.H. (1989) *The Primitive Edge of Experience.* (Jason Aronson, N.Y.)

Penn, William (1682) in *Quaker Faith & Practice* 23.02.

Rorty, Richard (2008) *Philosophy and the Mirror of Nature; 30th anniversary edition* (Princeton)

Segal, Hannah (1997) *Psychoanalysis, Literature & War.* (Routledge.)

Teilhard de Chardin (1961) *The Hymn of the Universe* (Harper & Rowe)

Todd, Emmanuel (1951) *The Explanation of Ideology: Family Structures & Social Systems.* (English trans. 1986. Blackwell)

Trollope, Joanna (1992) *The Rector's Wife* (Black Swan)

Wildwood, Alex. (2001) *Tradition & transition: opening to the Sacred yesterday & today* 'The Woodbrooke Journal' no. 9.

Williams, H.A. (1977) *BecomingWhat I Am* (Darton, Longman & Todd)

Williams, Rowan (2000) *Lost Icons* (T.&T. Clark)

Winnicott D.W. (1950) 'The Meaning of the Word "Democracy"' in *Home is Where We Start From* (Penguin, 1986)

Zeki, Semir (2009) *Splendours and Miseries of the Brain* (Wiley-Blackwell)

Index

Afghanistan 33, 56, 65, 66
aggression 17, 31, 48, 127, 130
alexithymia 38
anxiety 108, 111, 127
    existential 6, 106, 107, 155, 158, 159, 162, 163
Armstrong, K. 131, 155, 160
Arts and Crafts Movement 75
Art Deco 95
Augustine of Hippo 122
authority 18, 29, 46, 49, 50, 58
    -arianism 93
autistic-contiguous 86 - 88

bureaucratic centralism 59, 60, 62, 106, 152
biology 36, 59, 62, 125, 130, 158
Bion, W. R. 83, 167
bi-polar 12, 24, 37
birth 10, 16, 24, 25, 33, 34, 38, 105, 133, 143, 149
Bloomsbury group 77
Blue, L. 160
boundaries 47, 58, 69, 79, 82, 90, 92, 106, 107, 111, 112, 120, 134, 138, 140, 155

Caesarian section 24, 33
Celtic Church 122
Chasseguet-Smirgel, J. 82, 84
Christianity 13, 44, 121, 130, 152, 153, 154, 165, 167
Churchill, W. 76
climate change 26, 95, 164
communalism 60
Communism 28, 112, 162
corruption 86, 97, 101, 164
creation theology 91

death instinct 81, 147

debt 12, 23 - 24, 26, 77, 84, 91, 97, 98
de-centralisation 60
democracy 9, 18, 54, 56, 63, 64, 133, 167, 170
   -cratic 59, 133
denial 6, 12, 23, 48, 58, 59, 71, 80, 82, 85, 91, 92, 97, 112, 164, 166
   manic 51, 89
depression 15, 24, 28, 37, 39, 54, 79, 81, 89, 95, 97, 101, 136
   -ive 11, 12, 23, 38, 53, 80, 86, 87, 95, 97, 147
   Great 79, 95
devolution 55, 60
dynamic localisation 144

Eden 13, 25
empire 8, 28, 49, 53, 54, 67, 93, 97, 101, 128, 152, 153
Evangelical 71, 87, 93, 96

Fabian Society 74
facilitating environment 11, 58
fantasy 12, 52, 73, 99, 100, 102, 147, 164
Fascism 27-28
fathering 39, 40, 73, 96
feminism 83, 128
folie á deux 147
Fox, G. 110, 138, 139, 141
Freud, S. 10, 77, 88, 98, 114, 115, 142 - 144, 147, 150
Fromm-Reichmann, F. 38
fundamentalism 71, 111, 155, 163

Gestalt 12, 57, 91
Gospel of Thomas 140
grandiosity 49
greed 22, 89
Guantanamo Bay 54, 64

Hardy, A. 120, 137
homogenisation 82, 84

imperfections 119
Incarnation 13, 122, 131, 160
individualism 73, 89, 101
instrumentality 85, 100
International style 27, 80
internet 99 - 102
'Irish Problem' 75

Jung, C. G. 10, 114 - 120, 143, 149
justice 63, 133
    restorative 102

Kierkegaard, S. 140
Klein, M. 12, 37, 78
 -ian   71, 81, 147

latency 90, 148
Locke, J. 57, 61
Luria, A. 143

Manichean 121, 122, 130, 148, 163
Maxwell, J. C. 109
membrane 134, 135
    semi-permeable 134, 135
mother 12, 15, 16, 24, 33, 38, 40, 47, 50, 116, 129, 149
    -ing 40, 45, 73, 116
migration 28, 31, 67, 68, 111, 135, 164

Napoleon 9, 54, 126
narcissism 15, 89
New Labour 59, 60, 61, 71
Nonconformists 54, 76
Normans 68, 125, 126

Ogden, T. 86, 87
open-loop 51, 65

pacifism 124, 129
paranoid-schizoid 86 - 88, 93, 105
patriarchy 95
 -al   36
Penn, W. 124, 131, 139
perfect 11, 12, 13, 119, 131
    -ibility 9, 11, 12
    -ion 12
perverse 6, 82, 83, 85 - 86, 94, 88, 91 - 93, 97, 102, 112, 164
perversion 20, 58, 69, 78, 82, 84 - 86, 89, 100
Pill, the 5, 80
post-modernism 84
Poundbury 27, 30
projection 15, 16, 20

Quakers 114, 122, 123, 124, 140, 141, 156

reason 41, 42, 43, 47, 90, 109, 116, 118, 134, 137, 140, 149, 162
Reith, J. 84
revolution 28, 29, 93, 104, 143
   French 61
   industrial 54, 67, 74, 76, 77, 86

schizoid defence 121
scandal 17, 84, 97, 164
science 54, 84, 108, 109, 111, 113, 136, 154, 155
slavery 11, 18 - 20, 43
socialisation 69, 73, 98, 129, 148
Socialism 13, 56 - 62, 74, 101
society, multi-cultural 67, 68
Solms, K. K. & M. 144, 145, 146, 150
Spielrein, S. 143
spirituality 102, 114, 115, 121, 136, 151, 152
Strachey, A. 77, 79, 169
  J. 77
Swinging Sixties 51, 79, 80

Teilhard de Chardin 160
Todd, E. 43

uncertainty 106, 113, 118

war 29, 35, 54, 55, 63, 77, 79, 86, 89, 124, 126, 127, 164
Wildwood, A. 131
Williams, H. A. 119, 131, 140
 R.   13, 90
Winnicott, D. W. 9, 11, 133
Wordsworth, W. 120, 138

zeitgeist 58, 59, 71
Zeki, S. 144 - 150